# A Natural History
# of Human Thinking

# A Natural History
# of Human Thinking

*Michael Tomasello*

Harvard University Press

Cambridge, Massachusetts
London, England

First Harvard University Press paperback edition, 2018
First printing

*Library of Congress Cataloging-in-Publication Data*

Tomasello, Michael.
A natural history of human thinking / Michael Tomasello.
pages cm
Includes bibliographical references and index.
ISBN 978-0-674-72477-8 (hardcover : alk. paper) | ISBN 978-0-674-98683-1 (pbk.)
1. Cognition—Social aspects.  2. Evolutionary psychology.
3. Psychology, Comparative.  I. Title.
BF311.T6473 2014
153—dc23      2013020185

*For* Rita, Anya, Leo, *and* Chiara

# Contents

# Preface

This book is a sequel—or, better, a prequel—to *The Cultural Origins of Human Cognition* (Harvard University Press, 1999). But it also has a slightly different focus. In the 1999 book the question was what makes human cognition unique, and the answer was culture. Individual human beings develop uniquely powerful cognitive skills because they grow to maturity in the midst of all kinds of cultural artifacts and practices, including a conventional language, and of course they have the cultural learning skills necessary to master them. Individuals internalize the artifacts and practices they encounter, and these then serve to mediate all of their cognitive interactions with the world.

In the current book, the question is similar: what makes human thinking unique? And the answer is similar as well: human thinking is fundamentally cooperative. But this slightly different question and slightly different answer lead to a very different book. The 1999 book was clean and simple because the data we had comparing apes and humans were so sparse. We could thus say things like "Only humans understand others as intentional agents, and this enables human culture." But we now know that the picture is more complex than this. Great apes appear to know much more about others as intentional agents than previously believed, and still they do not have human-like culture or cognition. Based on much research reported here, the critical difference now seems to be that humans not only understand others as intentional agents but also put their heads together with others in acts of shared intentionality, including everything from concrete acts of collaborative problem solving to complex cultural institutions. The focus now is thus less on culture as a process of transmission and more on culture as a process of social coordination—and indeed, we argue here that modern human cultures were made possible by an earlier evolutionary step in which individuals made a living by coordinating with others in relatively simple acts of collaborative foraging.

The specific focus on thinking means that this book does not simply document that humans participate in shared intentionality in a way that their nearest primate relatives do not, which has been done elsewhere. Rather, in addition, it examines the underlying thinking processes involved. To describe the nature of these thinking processes—in particular, to distinguish human thinking from that of other apes—we must characterize its component processes of cognitive representation, inference, and self-monitoring. The *shared intentionality hypothesis* claims that all three of these components were transformed in two key steps during human evolution. In both cases, the transformation was part of a larger change of social interaction and organization in which humans were forced to adopt more cooperative lifeways. In order to survive and thrive, humans were forced, twice, to find new ways to coordinate their behavior with others in collaborative (and then cultural) activities and to coordinate their intentional states with others in cooperative (and then conventional) communication. And this transformed, twice, the way that humans think.

The writing of this book, as most others, was made possible by the support of many institutions and people. I would like to thank the University of Pittsburgh Center for Philosophy of Science (John Norton, director and seminar leader extraordinaire) for hosting me for one peaceful semester of concentrated writing in the spring of 2012. I especially benefited during this stay from Bob Brandom's generosity with his time and thoughts on many topics central to the current enterprise. I thank Celia Brownell at the Pitt Department of Psychology and Andy Norman at Carnegie Mellon for many useful discussions during this semester as well. The ensuing summer I benefited greatly from presenting the themes of the book to the SIAS Summer Institute titled *The Second Person: Comparative Perspectives*, organized in Berlin by Jim Conant and Sebastian Rödl. The book is better for all of these encounters.

With regard to the manuscript itself, I would like to thank Larry Barsalou, Mattia Galloti, Henrike Moll, and Marco Schmidt for reading various chapters and providing very useful feedback. Of special importance, Richard Moore and Hannes Rakoczy each read the entire manuscript at a fairly early stage and provided a number of trenchant comments and suggestions, regarding both content and presentation. Thanks also to Elizabeth Knoll and three anonymous reviewers at Harvard University Press for a number of helpful comments and criticisms on the penultimate draft.

Last and most important, I thank my wife, Rita Svetlova, for providing constant and detailed critical commentary and suggestions throughout. Many ideas were made clearer through discussion with her, and confusing passages were made clear, or at least clearer, by her literate eye.

# A Natural History
# of Human Thinking

# I

## The Shared Intentionality Hypothesis

Only cooperation constitutes a process that can produce reason.

—JEAN PIAGET, *SOCIOLOGICAL STUDIES*

Thinking would seem to be a completely solitary activity. And so it is for other animal species. But for humans, thinking is like a jazz musician improvising a novel riff in the privacy of his own room. It is a solitary activity all right, but on an instrument made by others for that general purpose, after years of playing with and learning from other practitioners, in a musical genre with a rich history of legendary riffs, for an imagined audience of jazz aficionados. Human thinking is individual improvisation enmeshed in a sociocultural matrix.

How did this novel form of socially infused thinking come to be, and how does it work? One set of classic theorists has emphasized the role of culture and its artifacts in making possible certain types of individual thinking. For example, Hegel (1807) argued that the social practices, institutions, and ideologies of a particular culture at a particular historical epoch constitute a necessary conceptual framework for individual human reason (see also Collingwood, 1946). Peirce (1931–1935) claimed more specifically that virtually all of humans' most sophisticated types of thinking, including most especially mathematics and formal logic, are possible only because individuals have available to them culturally created symbolic artifacts such as Arabic numerals and logical notation. Vygotsky (1978) emphasized that human children grow up in the midst of the tools and symbols of their culture, including especially the linguistic symbols that preorganize their worlds for them, and during ontogeny they internalize the use of these artifacts, leading to the kind of internal dialogue that is one prototype of human thinking (see also Bakhtin, 1981).

The other set of classic theorists has focused on the fundamental processes of social coordination that make human culture and language possible in the first place. Mead (1934) pointed out that when humans interact with one another, especially in communication, they are able to imagine themselves in the role of the other and to take the other's perspective on themselves. Piaget (1928) argued further that these role-taking and perspective-taking abilities—along with a cooperative attitude—not only make culture and language possible but also make possible reasoning in which individuals subordinate their own point of view to the normative standards of the group. And Wittgenstein (1955) explicated several different ways in which the appropriate use of a linguistic convention or cultural rule depends on a preexisting set of shared social practices and judgments ("forms of life"), which constitute the pragmatic infrastructure from which all uses of language and rules gain their interpersonal significance. These social infrastructure theorists, as we may call them, all share the belief that language and culture are only the "icing on the cake" of humans' ultrasocial ways of relating to the world cognitively.

Insightful as they were, all of these classic theorists were operating without several new pieces of the puzzle, both empirical and theoretical, that have emerged only in recent years. Empirically, one new finding is the surprisingly sophisticated cognitive abilities of nonhuman primates, which have been discovered mostly in the last few decades (for reviews, see Tomasello and Call, 1997; Call and Tomasello, 2008). Thus, great apes, as the closest living relatives of humans, already understand in human-like ways many aspects of their physical and social worlds, including the causal and intentional relations that structure those worlds. This means that many important aspects of human thinking derive not from humans' unique forms of sociality, culture, and language but, rather, from something like the individual problem-solving abilities of great apes in general.

Another new set of findings concern prelinguistic (or just linguistic) human infants, who have yet to partake fully of the culture and language around them. These still fledgling human beings nevertheless operate with some cognitive processes that great apes do not, enabling them to engage with others socially in some ways that great apes cannot, for example, via joint attention and cooperative communication (Tomasello et al., 2005). The fact that these precultural and prelinguistic creatures are already cognitively unique provides empirical support for the social infrastructure theorists' claim that important aspects of human thinking emanate not from culture and language per se but, rather, from some deeper and more primitive forms of uniquely human social engagement.

Theoretically, recent advances in the philosophy of action have provided powerful new ways of thinking about these deeper and more primitive forms of uniquely human social engagement. A small group of philosophers of action (e.g., Bratman, 1992; Searle, 1995; Gilbert, 1989; Tuomela, 2007) have investigated how humans put their heads together with others in acts of so-called shared intentionality, or "we" intentionality. When individuals participate with others in collaborative activities, together they form joint goals and joint attention, which then create individual roles and individual perspectives that must be coordinated within them (Moll and Tomasello, 2007). Moreover, there is a deep continuity between such concrete manifestations of joint action and attention and more abstract cultural practices and products such as cultural institutions, which are structured—indeed, created—by agreed-upon social conventions and norms (Tomasello, 2009). In general, humans are able to coordinate with others, in a way that other primates seemingly are not, to form a "we" that acts as a kind of plural agent to create everything from a collaborative hunting party to a cultural institution.

Further in this theoretical direction, as a specific form of human collaborative activity and shared intentionality, human cooperative communication involves a set of special intentional and inferential processes—first identified by Grice (1957, 1975) and since elaborated and amended by Sperber and Wilson (1996), Clark (1996), Levinson (2000), and Tomasello (2008). Human communicators conceptualize situations and entities via external communicative vehicles *for* other persons; these other persons then attempt to determine why the communicator thinks that these situations and entities will be relevant for them. This dialogic process involves not only skills and motivations for shared intentionality but also a number of complex and recursive inferences about others' intentions toward my intentional states. This unique form of communication—characteristic not just of mature language use but also of the prelinguistic gestural communication of human infants—presupposes both a shared conceptual framework between communicative partners (a.k.a. common conceptual ground) and an appreciation of those partners' individual intentions and perspectives within it.

These new empirical and theoretical advances enable us to construct a much more detailed account than was previously possible of the social dimensions of human cognition in general. Our focus in this book is on the social dimensions of human thinking in particular. Although humans and other animals solve many problems and make many decisions based on evolved intuitive heuristics (so-called system 1 processes), humans and at least some other animals also solve some problems and make some decisions by thinking (system 2 processes; e.g., Kahneman, 2011). A specific focus on thinking is useful because it restricts our topic to a single cognitive process, but one that involves several key components, especially (1) the ability to cognitively represent experiences to oneself "off-line"; (2) the ability to simulate or make inferences transforming these representations causally, intentionally, and/or logically; and (3) the ability to self-monitor and evaluate how these simulated experiences might lead to specific behavioral outcomes—and so to make a thoughtful behavioral decision.

It seems obvious that, compared with other animal species, humans think in special ways. But this difference is hard to characterize using traditional theories of human thinking since they presuppose key aspects of the process that are actually evolutionary achievements. These are precisely the social aspects of human thinking that are our primary focus here. Thus, although many animal species can cognitively represent situations and entities at least somewhat abstractly, only humans can conceptualize one and the same situation or entity under differing, even conflicting, social perspectives (leading ultimately to a sense of "objectivity"). Further, although many animals also make simple causal and intentional inferences about external events, only humans make socially recursive and self-reflective inferences about others' or their own intentional states. And, finally, although many animals monitor and evaluate their own actions with respect to instrumental success, only humans self-monitor and evaluate their own thinking with respect to the normative perspectives and standards ("reasons") of others or the group. These fundamentally social differences lead to an identifiably different type of thinking, what we may call, for the sake of brevity, *objective-reflective-normative thinking.*

In this book we attempt to reconstruct the evolutionary origins of this uniquely human objective-reflective-normative thinking. The *shared intentionality hypothesis* is that what created this unique type of thinking—its processes of representation, inference, and self-monitoring—were adaptations for dealing with problems of social coordination, specifically, problems presented by individuals' attempts to collaborate and communicate with others (to *co*-operate with others). Although humans' great ape ancestors were social beings, they lived mostly individualistic and competitive lives, and so their thinking was geared toward achieving individual goals. But early humans were at some point forced by ecological circumstances into more cooperative lifeways, and so their thinking became more directed toward figuring out ways to coordinate with others to achieve joint goals or even collective group goals. And this changed everything.

There were two key evolutionary steps. The first step, reflecting the focus of social infrastructure theorists such as Mead and Wittgenstein, involved the creation of a novel type of small-scale collaboration in human foraging. Participants in this collaborative foraging created socially shared joint goals and joint attention (common ground), which created the possibility of individual roles and perspectives within that ad hoc shared world or "form of life." To coordinate these newly created roles and perspectives, individuals evolved a new type of cooperative communication based on the natural gestures of pointing and pantomiming: one partner directed the attention or imagination of the other perspectivally and/or symbolically about something "relevant" to their joint activity, and then that partner made cooperative (recursive) inferences about what was intended. To self-monitor this process the communicator had to simulate ahead of time the recipient's likely inferences. Because the collaboration and communication at this point were between ad hoc pairs of individuals in the moment—based on purely second-personal social engagement between "I" and "you"—we may refer to all of this as *joint intentionality*. When put to use in thinking, joint intentionality comprises perspectival and symbolic representations, socially recursive inferences, and second-personal self-monitoring.

The second step, reflecting the focus of culture theorists such as Vygotsky and Bakhtin, came as human populations began growing in size and competing with one another. This competition meant that group life as a whole became one big collaborative activity, creating a much larger and more permanent shared world, that is to say, a culture. The resulting group-mindedness among all members of the cultural group (including in-group strangers) was based on a new ability to construct common *cultural* ground via collectively known cultural conventions, norms, and institutions. As part of this process, cooperative communication became conventionalized linguistic communication. In the context of cooperative argumentation in group decision making, linguistic conventions could be used to justify and make explicit one's reasons for an assertion within the framework of the group's norms of rationality. This meant that individuals now could reason "objectively" from the group's agent-neutral point of view ("from nowhere"). Because the collaboration and communication at this point were conventional, institutional, and normative, we may refer to all of this as *collective intentionality*. When put to use in thinking, collective intentionality comprises not just symbolic and perspectival representations but conventional and "objective" representations; not just recursive inferences but self-reflective and reasoned inferences; and not just second-personal self-monitoring but normative self-governance based on the culture's norms of rationality.

Importantly, this evolutionary scenario does not mean that humans today are hardwired to think in these new ways. A modern child raised on a desert island would not automatically construct fully human processes of thinking on its own. Quite the contrary. Children are born with adaptations for collaborating and communicating and learning from others in particular ways—evolution selects for adaptive *actions*. But it is only in actually exercising these skills in social interaction with others during ontogeny that children create new representational formats and new inferential reasoning possibilities as they internalize, in Vygotskian fashion, their coordinative interactions with others into thinking for the self. The result is a kind of cooperative cognition and thinking, not so much creating new skills as cooperativizing and collectivizing those of great apes in general.

And so let us tell a story, a natural history, of how human thinking came to be, beginning with our great ape ancestors, proceeding through some early humans who collaborated and communicated in species-unique ways, and ending with modern humans and their fundamentally cultural and linguistic ways of being.

# 2

## Individual Intentionality

Understanding consists in imagining the fact.

—LUDWIG WITTGENSTEIN, *THE BIG TYPESCRIPT*

Cognitive processes are a product of natural selection, but they are not its target. Indeed, natural selection cannot even "see" cognition; it can only "see" the effects of cognition in organizing and regulating overt actions (Piaget, 1971). In evolution, *being* smart counts for nothing if it does not lead to *acting* smart.

The two classic theories of animal behavior, behaviorism and ethology, both focused on overt actions, but they somehow forgot the cognition. Classical ethology had little or no interest in animal cognition, and classical behaviorism was downright hostile to the idea. Although contemporary instantiations of ethology and behaviorism take some account of cognitive processes, they provide no systematic theoretical accounts. Nor are any other modern approaches to the evolution of cognition sufficient for current purposes.

And so to begin this account of the evolutionary emergence of uniquely human thinking, we must first formulate, in broad outline, a theory of the evolution of cognition more generally. We may then begin our natural history proper by using this theoretical framework to characterize processes of cognition and thinking in modern-day great apes, as representative of humans' evolutionary starting point before they separated from other primates some six million years ago.

### Evolution of Cognition

All organisms possess some reflexive reactions that are organized linearly as stimulus-response linkages. Behaviorists think that all behavior is organized in this way, though in complex organisms the linkages may be learned and become associated with others in various ways. The alternative is to recognize that complex organisms also possess some adaptive specializations that are organized circularly, as feedback control systems, with built-in goal states and action possibilities. Starting from this foundation, cognition evolves not from a complexifying of stimulus-response linkages but, rather, from the individual organism gaining (1) powers of flexible decision-making and behavioral control in its various adaptive specializations, and (2) capacities for cognitively representing and making inferences from the casual and intentional relations structuring relevant events.

Adaptive specializations are organized as self-regulating systems, as are many physiological processes such as the homeostatic regulation of blood sugar and body temperature in mammals. These specializations go beyond reflexes in their capacity to produce adaptive behavior in a much wider range of circumstances, and indeed, they may be quite complex, for example, spiders spinning webs. There is no way that a spider can spin a web using only stimulus-response linkages. The process is too dynamic and dependent on local context. Instead, the spider must have goal states that it is motivated to bring about, and the ability to perceive and act so as to bring them about in a self-regulated manner. But adaptive specializations are still not cognitive (or only weakly cognitive) because they are unknowing and inflexible by definition: perceived situations and behavioral possibilities for goal attainment are mostly connected in an inflexible manner. The individual organism does not have the kind of causal or intentional understanding of the situation that would enable it to deal flexibly with "novel" situations. Natural selection has designed these adaptive specializations to work invariantly in "the same" situations as those encountered in the past, and so cleverness from the individual is not needed.

Cognition and thinking enter the picture when organisms live in less predictable worlds and natural selection crafts cognitive and decision making processes that empower the individual to recognize novel situations and to deal flexibly, on its own, with unpredictable exigencies. What enables effective handling of a novel situation is some understanding of the causal and/or intentional relations involved, which then suggests an appropriate and potentially novel behavioral response. For example, a chimpanzee might recognize that the only tool available to her in a given situation demands, based on the physical causality involved, manipulations she has never before performed toward this goal. A cognitively competent organism, then, operates as a control system with reference values or goals, capacities for attending to situations causally or intentionally "relevant" to these reference values or goals, and capacities for choosing actions that lead to the fulfillment of these reference values or goals (given the causal and/or intentional structure of the situation).

This description in control system terms is basically identical to the classic belief-desire model of rational action in philosophy: a goal or desire coupled with an epistemic connection to the world (e.g., a belief based on an understanding of the causal or intentional structure of the situation) creates an intention to act in a particular way.[1]

We will refer to this flexible, individually self-regulated, cognitive way of doing things as *individual intentionality*. Within this self-regulation model of individual intentionality, we may then say that thinking occurs when an organism attempts, on some particular occasion, to solve a problem, and so to meet its goal not by behaving overtly but, rather, by imagining what would happen if it tried different actions in a situation—or if different external forces entered the situation—before actually acting. This imagining is nothing more or less than the "off-line" simulation of potential perceptual experiences. To be able to think before acting in this way, then, the organism must possess the three prerequisites outlined above: (1) the ability to cognitively represent experiences to oneself "off-line," (2) the ability to simulate or make inferences transforming these representations causally, intentionally, and/or logically, and (3) the ability to self-monitor and evaluate how these simulated experiences might lead to specific behavioral outcomes—and so to make a thoughtful behavioral decision. The success or failure of a particular behavioral decision exposes the underlying processes of representation, simulation, and self-monitoring—indirectly, as it were—to the unrelenting sieve of natural selection.

### Cognitive Representation

Cognitive representation in a self-regulating, intentional system may be characterized both in terms of its content and in terms of its format. In terms of content, the claim here is that both the organism's internal goals and its externally directed attention (NB: not just perception but attention) have as content not punctate stimuli or sense data, but rather whole *situations*. Goals, values, and other reference values (pro-attitudes) are cognitive representations of situations that the organism is motivated to bring about or maintain. Although we sometimes speak of an object or location as someone's goal, this is really only a shorthand way of speaking; the goal is the situation of *having* the object or *reaching* the location. The philosopher Davidson (2001) writes, "Wants and desires are directed to propositional contents. What one wants is . . . *that* one has the apple in hand. . . . Similarly . . . someone who intends to go to the opera intends to make it the case *that* he is at the opera" (p. 126).

In this same manner, modern decision theory often speaks of the desire or preference *that* a particular state of affairs be realized.

If goals and values are represented as desired situations, then what the organism must attend to in its perceived environment is situations relevant to those goals and values. Desired situations and attended-to environmental situations are thus perforce in the same perceptually based, fact-like representational format, which enables their cognitive comparison. Of course, complex organisms also perceive less complex things, such as objects, properties, and events—and can attend to them for specific purposes—but in the current analysis they always do so as components of situations relevant to behavioral decision making.

To illustrate the point, let us suppose that the image in Figure 2.1 is what a chimpanzee sees as she approaches a tree while foraging.

FIGURE 2.1 What a chimpanzee sees

The chimpanzee perceives the scene in the same basic way that we would; our visual systems are similar enough that we see the same basic objects and their spatial relationships. But what situations does the chimpanzee attend to? Although she could potentially focus her attention on any of the potentially infinite situations that this image presents, at the current moment she must make a foraging decision, and so she attends to the situations or "facts" relevant to this behavioral decision, to wit (as described in English):

- that many bananas are in the tree
- that the bananas are ripe
- that no competitor chimpanzees are already in the tree
- that the bananas are reachable by climbing
- that no predators are nearby
- that escaping quickly from this tree will be difficult
- etc., etc.

For a foraging chimpanzee with the goal of obtaining food, given all of its perceptual and behavioral capacities and its knowledge of the local ecology, all of these are *relevant situations* for deciding what to do—all present in a single visual image and, of course, nonverbally. (NB: Even the absence of something expected, such as food not in its usual location, may be a relevant situation.)

Relevance is one of those occasion-sensitive judgments that cannot be given a general definition. But in broad strokes, organisms attend to situations as either (1) opportunities or (2) obstacles to the pursuit and maintenance of their goals and values (or as information relevant to predicting possible future opportunities or obstacles). Different species have different ways of life, of course, which means that they perceive or attend to different situations (and components of situations). Thus, for a leopard, the situation of bananas in a tree would not represent an opportunity to eat, but the presence of a chimpanzee would. For the chimpanzee, in contrast, the leopard's presence now presents an obstacle to its value of avoiding predators, and so it should look for a situation providing opportunities for escape, such as a tree to climb without low-hanging limbs—given its knowledge that leopards cannot climb such trees and its familiarity with its own tree-climbing prowess. If we now throw into the mix a worm resting on the banana's surface, the relevant situations for the three different species—the obstacles and opportunities for their respective goals—would overlap even less, if at all. Relevant situations are thus determined jointly by the organism's goals and values, its perceptual abilities and knowledge, and its behavioral capacities, that is to say, by its overall functioning as a self-regulating system. Identifying situations relevant for a behavioral decision thus involves an organism's whole way of life (von Uexküll, 1921).[2]

In terms of representational format, the key is that to make creative inferences that go beyond particular experiences, the organism must represent its experiences as types, that is to say, in some generalized, schematized, or abstract form. One plausible hypothesis is a kind of exemplar model in which the individual in some sense "saves" the particular situations and components to which it has attended (many models of knowledge representation have attention as the gateway). There is then generalization or abstraction across these in a process that we might call *schematization*. (Langacker's [1987] metaphor is of a stack of transparencies, each depicting a single situation or entity, and schematization is the process of looking down through them for overlap.) We might think of the result of this process of schematization as cognitive models of various types of situations and entities, for example, categories of objects, schemas of events, and models of situations. Recognizing a situation or entity as a token of a known type—as an exemplar of a cognitive category, schema, or model—enables novel inferences about the token appropriate to the type.

Categories, schemas, and models as cognitive types are nothing more or less than imagistic or iconic schematizations of the organism's (or, in some cases, its species') previous experience (Barsalou, 1999, 2008). As such, they do not suffer from the indeterminacy of interpretation that some theorists attribute to iconic representations considered as mental pictures, that is, the indeterminacy of whether this image is of a banana, a fruit, an object, and so forth (Crane, 2003). They do not because they are composed of individual experiences in which the organism was attending to a relevant (already "interpreted") situation. Thus, the organism "interprets," or understands, particular situations and entities in the context of its goals as it assimilates them to known (cognitively represented) types: "This is another one of those."

### Simulation and Inference

Thinking in an organism with individual intentionality involves simulations or inferences that connect cognitive representations of situations and their components in various ways. First are those instrumental inferences that occur in behavioral decision making of the "what would happen if . . ." variety. For example, in a concrete problem-solving situation—such as a rock preventing the movement of a stick beneath it—some organisms might go through the kind of inferential simulation that Piaget (1952) called "mental trial and error": the organism imagines a potential action and its consequences. Thus, a chimpanzee might simulate imaginatively what would happen if she forcefully tugged at the stick, without actually doing it. If she judged that this would be futile given the size and weight of the rock, she might decide to push the rock aside before pulling the stick.

Also possible are inferences about causal and intentional relations created by outside forces and how these might affect the attainment of goals and values. For example, a chimpanzee might see a monkey feeding in the banana tree and infer that there are no leopards nearby (because if there were, the monkey would be fleeing). Or, upon finding a fig on the ground, a bonobo might infer that it will have a sweet taste and that there is a seed inside—based on a categorization of the encountered fig as "another one of those" and the natural inference that this one will have the same properties as others in the category. Or an orangutan might recognize a conspecific climbing a tree as an intentional event of a particular type, and then infer something about goals and attention as intentional causes and so predict the climber's impending actions. Schematizing over such experiences (again aided, perhaps, by the experience of the species), individuals may potentially build cognitive models of general patterns of causality and intentionality.

The best way to conceptualize such processes is in terms of off-line, image-based simulations, including novel combinations of represented events and entities that the individual herself has never before directly experienced as such, for example, an ape imagining what the monkey would do if a leopard entered the scene (see Barsalou, 1999, 2008, for relevant data on humans; Barsalou, 2005, extends the analysis to nonhuman primates). Importantly, the combinatorial processes themselves will include causal and intentional relations that connect different real and imagined situations, and also "logical" operations such as the conditional, "negation," exclusion, and the like. These logical operations are not themselves imagistic cognitive representations but, rather, cognitive procedures (enactive, in Bruner's terms, or operative, in Piaget's terms) that the organism accesses only through actual use. Concrete examples of how this works will be given when we look more closely at great ape thinking in the section that follows.

*Behavioral Self-Monitoring*

To think effectively, an organism with individual intentionality must be able to observe the outcome of its actions in a given situation and evaluate whether they match the desired goal state or outcome. Engaging in some such processes of behavioral self-monitoring and evaluation is what enables learning from experience over time.

A cognitive version of such self-monitoring enables the agent, as noted above, to inferentially simulate a potential action-outcome sequence ahead of time and observe it—as if it were an actual action-outcome sequence—and then evaluate the imagined outcome. This process creates more thoughtful decision making through the precorrection of errors. (Dennett [1995] calls it Popperian learning because failure means that my hypothesis "dies," not me.) For example, consider a squirrel on one tree branch gearing up to jump to another. One can see the muscles preparing, but in some cases the squirrel decides the leap is too far and so, after feigning some jumps, climbs down the trunk and then back up the other branch. The most straightforward description of this event is that the squirrel is observing and evaluating a simulation of what it would experience if it leaped; for example, it would experience missing the branch and falling—a decidedly negative outcome. The squirrel must then use that simulation to make a decision about whether to actually leap. Okrent (2007) holds that imagining the possible outcomes of different behavioral choices ahead of time, and then evaluating and deciding for the one with the best imagined outcome, is the essence of instrumental rationality.

This kind of self-monitoring, requiring what some call executive functioning, is cognitive because the individual, in some sense, observes not just its actions and their results in the environment but also its own internal simulations. It is also possible for the organism to assess things like the information it has available for making a decision in order to predict the likelihood that it will make a successful choice (before it actually chooses). Humans even use the imagined evaluations of other persons—or the imagined comprehension of others in the case of communication—to evaluate potential behavioral decisions. Whatever its specific form, internal self-monitoring of some kind is critical to anything we would want to call thinking, as it constitutes, in some sense, the individual knowing what it is doing.

## Thinking like an Ape

We begin our natural history of the evolutionary emergence of uniquely human thinking with a focus on the last common ancestor of humans and other extant primates. Our best living models for this creature are humans' closest primate relatives, the nonhuman great apes (hereafter, great apes), comprising chimpanzees, bonobos, gorillas, and orangutans—especially chimpanzees and bonobos, who diverged from humans most recently, around 6 million years ago. When cognitive abilities are similar among the four species of great ape but different in humans, we presume that the apes have conserved their skills from the last common ancestor (or before) whereas humans have evolved something new.

Our characterizations of the cognitive skills of this last common ancestor will derive from empirical research with great apes, cast in the theoretical framework of individual intentionality just elaborated: behavioral self-regulation involving cognitive models and instrumental inferences, along with some form of behavioral self-monitoring. Because humans share with other apes such a recent evolutionary history—along with the same basic bodies, sense organs, emotions, and brain organization—in the absence of evidence our default assumption will be evolutionary continuity (de Waal, 1999). That is to say, when great apes behave identically with humans, especially in carefully controlled experiments, we will assume continuity in the underlying cognitive processes involved. The onus of explanation is thus on those who posit evolutionary discontinuities, a challenge we embrace in later chapters.

### Great Apes Think about the Physical World

Processes of great ape cognition and thinking may be usefully divided into those concerning the physical world, structured by an understanding of physical causality, and those concerning the social world, structured by an understanding of agentive causality, or intentionality. Primate cognition of the physical world evolved mainly in the context of foraging for food (see Tomasello and Call, 1997, for this theoretical claim and supporting evidence); this is thus its "proper function" (in Millikan's [1987] sense). In order to procure their daily sustenance, primates (as mammals in general) evolved the proximate goals, representations, and inferences for (1) finding food (requiring skills of spatial navigation and object tracking), (2) recognizing and categorizing food (requiring skills of feature recognition and categorization), (3) quantifying food (requiring skills of quantification), and (4) procuring or extracting food (requiring skills of causal understanding). In these most basic skills of physical cognition, all nonhuman primates would seem to be generally similar (Tomasello and Call, 1997; Schmitt et al., 2012).

What great apes are especially skillful at, compared with other primates, is tool use—which one might characterize as not just understanding causes but actually manipulating them. Other primates are mostly not skilled tool users at all, and when they are it is typically in only one fairly narrow context (e.g., Fragaszy et al., 2004). In contrast, all four species of great ape are highly skilled at using a variety of tools quite flexibly, including using two tools in succession in a task, using one tool to rake in another (which is then needed to procure food), and so forth (Herrmann et al., 2008). Classically, tool use is thought to require the individual to assess the causal effect of its tool manipulations on the goal object or event (Piaget, 1952), and so the flexibility and alacrity with which great apes succeed in using novel tools suggest that they have one or more general cognitive models of causality guiding their use of these novel tools.

Great apes' skills with manipulating causal relations via tools may be combined in interesting ways with processes of cognitive representation and inference. For example, Marín Manrique et al. (2010) presented chimpanzees with a food extraction problem that they had never before seen. Its solution required a tool with particular properties (e.g., it had to be rigid and of a certain length). The trick was that the potential tools they could use were in a different room, out of sight of the problem. To solve this task, individuals had to first comprehend the causal structure of the novel problem, and then keep that structure cognitively represented while approaching and choosing a tool in the other room. Many individuals did this, often from the first trial onward, suggesting that they assimilated the novel problem to a known cognitive model having a certain causal structure, which they then kept with them as they entered the adjoining room. They then simulated the use of at least some of the available tools and the likely outcome in each case through the medium of this cognitive model—before actually choosing a tool overtly. In the study of Mulcahy and Call (2006), bonobos even saved a tool for future use, presumably imagining the future situation in which they would need it.

The simulations or inferences involved here have logical structure. This is not the structure of formal logic but, rather, a structure based on causal inferences. The idea is that causal inferences have a basic if-then logic and so lead to "necessary" conclusions: if A happens, then B happens (because A caused B). Bermudez (2003) calls inferences of this type protoconditional because the necessity is not a formal one but a causal one. In the experiment of Marín Manrique et al. (2010), as an ape simulates using the different tools, she infers "if a tool with property A is used, then B must happen." One thus gets a kind of proto-*modus ponens* by then actually using the tool with property A in the expectation that B will indeed happen as a causal result (if A happens, then B happens; A happens; therefore B will happen). This is basically a forward-facing inference, from premise or cause to conclusion or effect.

In another set of recent experiments, we can see backward-facing inferences, that is, from effect to cause. Call (2004) showed chimpanzees a piece of food, which was then hidden in one of two cups (they did not know which). Then, depending on condition, the experimenter shook one of the cups. The relevant background knowledge for success in this experiment is as follows: (1) the food is in one of the two cups (learned in pretraining), and (2) shaking the cup with food will result in noise, whereas shaking the cup without food will result in silence (causal knowledge brought to the experiment). The two conditions are shown in Figure 2.2, using iconic representations to depict something of the way the apes understand the situation.

(The iconic diagrams modeling great ape cognitive representations in Figure 2.2 are not uninterpreted pictures but symbols in a theoretical metalanguage that mean what we agree that they mean. So they are meant to depict the ape's interpreted experience when she has seen the cup as a cup and the noise as coming from the cup, and so on. Importantly, these diagrams are created within the confines of a restrictive theory of the possibilities of great ape cognition. Following Tomasello's (1992) depictions for one-year-old human children, we make the diagrams out of concrete spatial-temporal-causal elements that may be posited to be a part of the apes' cognitive abilities based on empirical research. Then, the logical structure—based on the protoconditional and protonegation—is posited to be necessary to explain apes' actions in specific experimental situations. The logical operations are depicted in English words, since the ape does not have perception-based representations of them, but only procedural competence with them.)

In condition 1, an experimenter shook the cup with food. In this case the chimpanzee observed a noise being made and had to infer backward in the causal chain to what might have caused it, specifically, the food hitting the inside of the cup. This is a kind of abduction (not logically valid, but an "inference to best explanation"). That is, (1) the shaking cup is making noise;

**BACKGROUND KNOWLEDGE:**

a)

b)

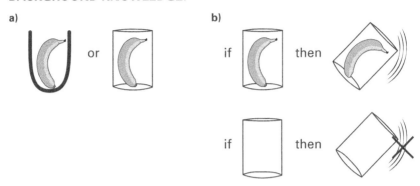

**CONDITION 1:**

Observation:      Prediction/ Inference:

**Inference to
Best Explanation**

**CONDITION 2:**

Observation:      Prediction:      Prediction:

**Proto-Disjunctive
Syllogism**

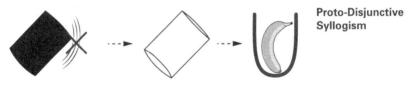

FIGURE 2.2 Ape inferences in finding hidden food (Call, 2004)

(2) if the food were inside the shaking cup, then it would make noise; (3) therefore, the food is inside the cup. In condition 2, the experimenter shook the empty cup. In this case the chimpanzee observed only silence and had to infer backward in the causal chain to why that might be, specifically, that there was no food in the cup. This is a kind of proto-*modus tollens*: (1) the shaking cup is silent; (2) if the food were inside the shaking cup, then it would make noise; (3) therefore, the food must not be in the cup (the shaken cup must be empty). The chimpanzees made this inference, but they also made an additional one. They combined their understanding of the causality of noise making in this context with their preexisting knowledge that the food was in one of the two cups to locate the food in the *other*, nonshaken cup (if the food is not in this one, then it must be in that one; see bottom row in Figure 2.2). This inferential paradigm thus involves the kind of exclusion inference characteristic of a disjunctive syllogism.

Negation is a very complex cognitive operation, and one could easily object to the use of negation in these proposed accounts of great ape logical inferences. But Bermudez (2003) makes a novel theoretical proposal about some possible evolutionary precursors to formal negation that make these accounts much more plausible. The proposal is to think of a kind of protonegation as simply comprising exclusionary opposites on a scale (contraries), such as presence-absence, noise-silence, safety-danger, success-failure, and available-not available. If we assume that great apes understand polar opposites such as these as indeed mutually exclusive—for example, if something is absent, it cannot be present, or if it makes noise it cannot be silent—then this could be a much simpler basis for the negation operation. All of the current descriptions assume protonegation of this type.

When taken together, the conditional (if-then) and negation operations structure all of the most basic paradigms of human logical reasoning. The claim is thus that great apes can solve complex and novel physical problems by assimilating key aspects of the problem situation to already known cognitive models with causal structure and then use those models to simulate or make inferences about what has happened previously or what might happen next—employing both a kind of protoconditional and a kind of protonegation in both forward-facing and backward-facing paradigms. Our general conclusion is thus that since the great apes in these studies are using cognitive models containing general principles of causality, and they are also simulating or making inferences in various kinds of protological paradigms, with various kinds of self-monitoring along the way, what the great apes are doing in these studies is thinking.

### Great Apes Think about the Social World

Primate cognition of the social world evolved mainly in the context of competition within the social group for food, mates, and other valued resources (see Tomasello and Call, 1997); competitive social interactions are thus its "proper function." In order to outcompete groupmates, individual primates evolved the proximate goals, representations, and inferences for (1) recognizing individuals in their social group and forming dominance and affiliative relationships with them and (2) recognizing third parties' social relationships with one another, such as parent or dominant or friend, and taking these into account. These abilities enable individuals to better predict the behavior of others in a complex "social field" (Kummer, 1972). Despite important species differences of social structure and interaction, in these most basic of skills of social cognition all primates would appear to be generally similar (e.g., see Tomasello and Call, 1997; see also the chapters in Mitani et al., 2012).

Beyond recognizing social relationships based on observed social interactions, great apes also understand that other individuals have goal-situations that they are pursuing and perceived-situations in the environment that they are attending to—so that together the individual's goals and perceptions (and her assessment of any relevant obstacles and opportunities for goal achievement in the environment) determine her behavior. This means that nonhuman great apes not only are intentional agents themselves but also understand others as intentional agents (i.e., as possessing individual intentionality; Call and Tomasello, 2008).

Consider the following experiment. Hare et al. (2000) had a dominant and subordinate chimpanzee compete for food in a novel situation in which one piece of food was out in the open and one piece of food was on the subordinate's side of a barrier where only she could see it. In this situation the subordinate knew that the dominant could see the piece out in the open and so he would go for it as soon as he could, whereas he could not see the other piece (i.e., he saw the barrier only) and so would not go for that piece (i.e., he would stay on his side). When her door was opened (slightly before the dominant's), the subordinate chose to pursue the food on her side of the barrier; she knew what the dominant could and could not see. In an important variation, subordinate chimpanzees avoided going for food that a dominant could not see now but had seen hidden in one of the locations some moments before; they knew that he *knew* where the hidden food was located (Hare et al., 2001; Kaminski et al., 2008). In still another variation, in a back-and-forth foraging game, chimpanzees knew that if their competitor chose first, he would choose a board that was lying slanted on the table (as if something were underneath) rather than a flat board (under which there could be nothing); they knew what kind of inference he would make in the situation (Schmelz et al., 2011). Chimpanzees thus know that others see things, know things, and make inferences about things.

But beyond exploiting their understanding of what others do and do not experience and how this affects their behavior, great apes sometimes even attempt to manipulate what others experience. In a series of experiments, Hare et al. (2006) and Melis et al. (2006a) had chimpanzees compete with a human (sitting in a booth-like apparatus) for two pieces of food. In some conditions, the human could see the ape equally well if it approached either piece of food; in these cases, the apes had no preference for either piece. But in the key condition, a barrier was in place so that the apes could approach one piece of food without being seen—which is exactly what they did. They even did this when they themselves could not see the human in either case. (They had to choose to reach for food from behind a barrier in both cases, but through a clear tunnel in one case and an opaque one in the other.) Perhaps most impressive, the same individuals also preferentially chose to pursue food that they could approach silently—so that the distracted human competitor did not know they were doing so—as opposed to food that required them to make noise en route. This generalization to a completely different perceptual modality speaks to the power and flexibility of the cognitive models and inferences involved.

Importantly analogous to the domain of physical cognition, the chimpanzees in these studies not only made productive inferences based on a general understanding of intentionality but also connected their inferences into paradigms to both predict and even manipulate what others would do (see Figure 2.3). The background knowledge required in all of these food competition experiments is that a competitor will go for a piece of food if and only if (1) he has the goal of having it and (2) he perceives its location (e.g., at location A). The protoconditional inferences in the Hare et al. (2000) experiment follow straightforwardly from this: if the dominant wants the banana and sees it at location A, then she will go to location A. Also analogous to the domain of physical cognition, in these food competition experiments chimpanzees make use of protonegation. Thus, conceptualizing protonegation in terms of polar opposites, in the Hare et al. experiments chimpanzees know that if the competitor sees only the barrier, then she will stay in place (i.e., if she does not see the food she will not go for it; see Figure 2.3, condition C). In the Melis et al. (2006a) concealment experiments, chimpanzees also understand that if the human sees only the barrier, or hears only silence, she will remain sitting peacefully (i.e., if she does not see or hear my approach, she will not reach for the food), and so they approach such that the other sees only barrier or hears only silence.[3]

And so, just as in the domain of physical cognition, in the domain of social cognition what great apes are especially good at is manipulation. This special facility for social manipulation also comes out clearly in their gestural communication. (Their vocal communication is mostly hardwired, and similar to that of monkeys, so not of great interest for the question of thinking.) All four species of great apes communicate with others gesturally in ways that nonape primates mostly do not. They ritualize from their social interactions with others certain intention-movements—such as raising the arm to begin play-hitting as an instigation to play—that they then use flexibly to manipulate the behavior of others. Perhaps even more important, they also use a number of attention-getting gestures—for example, slapping the ground to get others to look at them—in order to manipulate the attention of others. And they even adopt with humans something like a ritualized reaching or pointing gesture, one that is not in their natural repertoire with conspecifics, demonstrating especially clearly the flexibility of great apes' skills for manipulating the behavior and attention of others (see Call and Tomasello, 2007, for a review). Great apes' gestural communication thus shows again their special skills at manipulating causes.

A final experiment demonstrates something like backward-facing inferences in the social domain. Buttelmann et al. (2007) tested six human-raised chimpanzees in the so-called rational imitation paradigm of Gergely et al. (2002). Individuals saw a human perform an unusual action on an apparatus to produce an interesting result. The idea was that in one condition the physical constraints of the situation forced the human to use an unusual action; for example, he had to turn on a light with his head because his hands were occupied holding a blanket, or he had to activate a music box with his foot because his hands were occupied with a stack of books. When given their turn with the apparatus and no constraints in effect, the chimpanzees dis-

## BACKGROUND KNOWLEDGE:

if                    then

## FORWARD INFERENCE:

**Observation:**                    **Prediction/ Inference:**

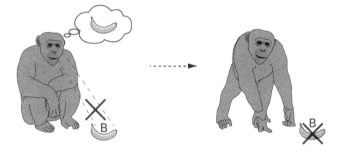

## BACKWARD INFERENCE:

**Observation:**                    **Prediction/ Inference**

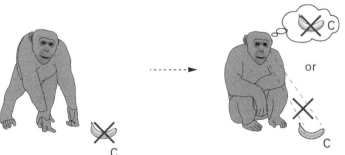

FIGURE 2.3 Ape inferences in competing for food (Hare et al., 2000)

counted the unusual action and used their hands as they normally would. However, when they saw the human use the unusual action when there was no physical constraint dictating this—he just turned on the light with his head for no discernable reason—they quite often copied the unusual behavior themselves. The most natural interpretation of this differentiated pattern of response would be that the apes employed a kind of proto-*modus tollens*, from effect to cause with protonegation, similar to that in the Call (2004) shaking cups study: (1) he is not using his hands; (2) if he had a free choice, he would be using his hands; (3) therefore he must not have a free choice (in one case for obvious reasons; in the other not).

These studies demonstrate that great apes can solve complex social problems, just as they solve complex physical problems, by assimilating key aspects of the problem situation to a cognitive model—which in this case embodies a general understanding of intentionality—and then using that model to simulate or make inferences about what has happened or what might happen next. Great apes employ both a kind of protoconditional and a kind of protonegation—in both forward-facing and backward-facing modes—in the context of protological paradigms of social inferring. Our conclusion is thus that in the social domain, as well as the physical domain, what the great apes in these studies are doing is thinking.

### Cognitive Self-Monitoring

Great apes in these studies are clearly not just automatically flipping through behavioral alternatives and reacting to a goal match; they monitor, and so in some sense know, what they are doing in order to make more effective decisions. On the level of action (recall the hesitant squirrel), recent studies of great apes have shown that they can (1) delay taking a smaller reward so as to get a larger reward later, (2) inhibit a previously successful response in favor of a new one demanded by a changed situation, (3) make themselves do something unpleasant for a desired reward at the end, (4) persist through failures, and (5) concentrate through distractions. Specifically, in a comprehensive comparative study, chimpanzees' ability to do these things was roughly comparable to that of three-year-old human children (Herrmann et al., submitted). These are all skills referred to variously as impulse control, attentional control, emotion regulation, and executive function—though we prefer to use the terms *behavioral self-monitoring*, for more action-based self-regulation, and *cognitive self-monitoring* (and in some cases, *self-reflection*), for more cognitive versions of the process.

Evidence that apes go beyond just behavioral self-monitoring and engage in cognitive self-monitoring comes from several experimental paradigms used with nonhuman primates (sometimes referred to as studies of metacognition). In the best-known paradigm, employed mostly with rhesus monkeys, individuals must make a discrimination (or remember something) to get a highly desirable reward. But if they fail to make the discrimination or remember correctly, they get nothing and must take a time-out before the next trial. The trick is that on each trial individuals have the possibility of opting out of the problem and going for a lesser reward with 100% certainty, which also means no time-out before the next trial. Many individuals thus develop a strategy of opting out of only those discrimination or memory tasks that they are especially likely to fail (Hampton, 2001). They seem to know that they do not know or that they do not remember.

In another paradigm involving chimpanzees, individuals either do or do not see the process of food being hidden in one of several tubes. When they see the hiding process, they choose a tube directly. When they do not see the hiding process, they go to some trouble to look into the tubes and discover where the food is located before choosing. Again, the apes seem to know when they do not know, or at least when they are uncertain, and seek to do something about it. Interesting for interpretation, variables that affect this process in apes are the same as those that affect it in humans: they are more likely to seek extra information if the reward is highly valuable or if it has been longer since they acquired the information (Call, 2010). Thus, as they are assessing situations and deciding what to do, if apes self-monitor and find that they have insufficient information to make an effective decision, this prompts them to gather information as prerequisite to their choice.

The interpretation of these experiments is not totally straightforward, but the apes clearly are doing some kind of self-monitoring and evaluation, just as they do in all of their intelligent decision making. What is new here is that they seem to be monitoring not just imagined actions and their imagined results, or imagined causes and their imagined outcomes, but also their own knowledge or memory—which they then use to make inferences about their likelihood of behavioral success. Great apes and other primates thus have some kind of access, at least in instrumental contexts, to their own psychological states. And even if this is not the fully human version of self-reflection (as we will argue later that it is not—because it lacks a social/perspectival dimension), this ability adds further to the conclusion that great apes are skillful with all three of the key components—abstract cognitive representations (models), protological inferential paradigms, and psychological self-monitoring and evaluation—that constitute what can only be called thinking.

## Cognition for Competition

Many theorists continue to maintain a kind of Cartesian picture of the difference between humans and other animals: humans have rational thinking, whereas other animals, including great apes, are simply stimulus-response machines with no ratiocination whatever. This view is held not only by behaviorist psychologists but also by many otherwise very thoughtful philosophers and cognitive scientists. But it is a factually incorrect view, in our opinion, and one that is grounded in an erroneous theory of the evolution of cognition (Darwin, 1859, 1871). Cognitive evolution does not proceed from simple associations to complex cognition, but rather from inflexible adaptive specializations of varying complexities to flexible, individually self-regulated intentional actions underlain by cognitive representations, inferences, and self-monitoring. The empirical research reviewed here (there is much more) clearly demonstrates, in our view, that great apes operate in this flexible, intelligent, self-regulated way—and they do so without language, culture, or any other forms of human-like sociality.

This is not to say, of course, that the interpretations of the studies cited here are the only possible ones. Thus, some theorists would contest the conclusion that great apes understand causal and intentional relations (e.g., Povinelli, 2000; Penn et al., 2008), proposing instead that they operate with some kind of noncognitive "behavioral rules." Others would propose that instead of causal, intentional, and logical inferences, great apes—just like rats and pigeons—operate only with associations (e.g., Heyes, 2005). And skepticism about cognitive self-monitoring in apes and other animals abounds (e.g., Carruthers and Ritchie, 2012). But behavioral rules (whose nature and origins have never been specified) cannot account for the flexibility with which great apes solve novel physical and social problems (Tomasello and Call, 2006), and association learning takes many dozens of trials to be effective, and this does not accord with the speed and flexibility with which great apes solve novel physical and social problems in experiments (Call, 2006). Although the empirical data are less clear-cut in the case of cognitive self-monitoring, Call's (2010) finding that the same factors affect the process in humans and great apes is highly suggestive that—in concrete situations, at least—the apes are genuinely self-monitoring the decision-making process.

In any case, our natural history of human thinking begins with this possibly somewhat generous account of great ape thinking. To summarize, thinking comprises three key components, and great apes operate in cognitively sophisticated ways with each of them.

## Schematic Cognitive Representations

First is the ability to work with some kind of abstract cognitive representations, to which the individual assimilates particular experiences. According to the best available evidence, great ape abstract cognitive representations— categories, schemas, and models—have three main features.

IMAGISTIC. Great ape cognitive representations are iconic or imagistic in format, based on processes of perceptual and motor experience (for the proposal that human infant cognitive representations are also iconic in format, see Carey, 2009; Mandler, 2012). It is difficult to imagine what else they could be.

SCHEMATIC. Great apes' imagistic representations are generalized or abstract: schematizations of the organism's perceptual experience of exemplar situations or entities (i.e., they have type-token structure). Importantly, iconic or imagistic schematizations are not uninterpreted "pictures" but, rather, amalgams of already understood (made relevant to existing cognitive models) exemplars. Thus, when Wittgenstein is searching for what could underlie our most basic processes of understanding, he speculates "imagining the fact": representing to ourselves the more general and already meaningful cognitive model to which the current situation is best assimilated as exemplar. Such cognitive models are already meaningful because a schematized and generalized understanding of causality and/or intentionality is part and parcel of many of apes' cognitive models of situations.

SITUATIONAL CONTENT. Great ape cognitive representations have as their most basic content situations, specifically, situations that are relevant to the individual's goals and values (e.g., that food is present or that a predator is absent). Obviously, representational content structured as whole situations prefigures, in some important sense, human propositional content (though it is not there yet). Apes can also, for specific purposes, schematize their experience with components of situations such as objects and events (e.g., a category of "figs").

### Causal and Intentional Inferences

The second key component is the ability to make inferences from cognitive representations. Great apes use their cognitive categories, schemas, and models productively to imagine or infer nonactual situations. These inferences have two main characteristics.

CAUSALLY AND INTENTIONALLY LOGICAL. Great ape inferences are based on their general understanding of causality and intentionality; they are causal and intentional inferences. But importantly, they still have logical structure—they form paradigms—based on facility with a kind of protoconditional (inferences between cause and effect in both the physical world and social world) and a kind of protonegation (based on mutually exclusive polar opposite or contraries, such as presence-absence). Apes thus possess protoversions of everything from modus tollens to disjunctive syllogism.

PRODUCTIVE. Great apes' cognitive representations and inferences are productive or generative in that they can support off-line simulations in which the subject infers or imagines nonactual situations (Barsalou, 1999, 2008). Nevertheless, some theorists might still doubt whether great ape thinking meets Evans's (1982) generality constraint. In this linguistically inspired account, each potential subject of a thought (or sentence) may be combined with multiple predicates, and each potential predicate may have multiple subjects. To do this nonlinguistically, an individual must be able not only to relate represented situations to one another but also to extract their components and use them in productive combinations to imagine novel situations.

With respect to a particular agent doing multiple things, great apes know, for example, that this leopard does lots of things like climb trees, eat chimpanzees, drink water, and so on. Indirect evidence for this is the fact that great apes pass object permanence tasks in which they must understand that the same object is going different places and doing different things (Call, 2001), and also that they can predict what particular individuals will do in situations based on their past experience with them (Hare et al., 2001). Further evidence is the fact that in experiments great apes individuate objects, so that if they see a particular object go behind a screen, they expect to find that particular object there, and if they see it leave and another replace it, they do not expect to find it there anymore—and if two identical objects go behind the screen they expect to find two objects. They are not "feature placing," but rather, they are tracking the self-same object or objects engaging in different actions across time (Mendes et al., 2008).

With respect to different individuals doing "the same thing," great apes know such individual things as leopards climb trees, snakes climb trees, monkeys climb trees—each in their own way. Here things are a bit more difficult evidentially because there are few if any nonverbal methods for investigating event schemas like *climbing*. But one hypothesis is that a nonverbal way of establishing an event schema is imitation. That is, an individual who imitates another knows at the very least that a demonstrator is doing X and then they themselves can do X—"the same thing"—as well (and perhaps other actors also). Although imitation is not their frontline strategy for social learning, great apes (at least those raised by humans) are nevertheless capable of reproducing the actions of others with some facility in some contexts (e.g., Tomasello et al., 1993; Custance et al., 1995; Buttelmann et al., 2007). Some apes also know when another individual is imitating them, again suggesting at least a rudimentary understanding of self-other equivalence (Haun and Call, 2008). But imitation involves just self and other. Since apes understand the goals of all agents, an alternative hypothesis might be that apes schematize acts of climbing based not on movements but on an understanding that the actor has the goal of getting up the tree—and that goal (not actions per se) provides the basis for an event schema across all individuals, with or without the self.

Great ape cognition thus goes at least some way toward meeting the generality constraint, although productivity may be limited. The claim would be that great ape productive thinking enables an individual to imagine, for example, that if I chase this novel animal it might climb a tree, even if I have never before seen this animal climb a tree. On the other hand, it may be that an ape could not imagine something contrary to fact (i.e., contrary to its causal understanding), such as a leopard flying, as humans are able to do with the aid of external communicative vehicles. Apes' sense of self-other equivalence may also be limited by the fact that imitation takes place sequentially, whereas much better for establishing self-other equivalence are situations in which the equivalence manifest in a single social interaction simultaneously (e.g., role reversal in the collaborative activities of humans).

### Behavioral Self-Monitoring

The third key component of great ape thinking is the ability to self-monitor the decision-making process. Many animal species self-monitor, and even anticipate, the outcomes of their behavioral decisions in the world. But great apes do more than this simple behavioral self-monitoring.

COGNITIVE SELF-MONITORING. Great apes (and some other primate species) also know when they have insufficient information to make an informed behavioral decision. As noted, monitoring outcomes is a basic prerequisite of a self-regulating system, and monitoring simulated outcomes is a characteristic of a cognitive system capable of thinking before it chooses. But monitoring the elements of the decision-making process itself—one's memory or powers of discrimination or the environmental information available—is a still further enrichment. Self-monitoring of this type implies some kind of "executive" oversight of the decision-making process itself.

And so we may imagine a common ancestor to humans and other great apes. Its daily life was like that of extant nonhuman apes: most waking hours spent in small bands foraging individually for fruit and other vegetation, with various kinds of social interactions, mostly competitive, interspersed. Our hypothesis is that this creature—and also probably australopithecines for the ensuing 4 million years of the human lineage—was individually intentional and instrumentally rational. It cognitively represented its physical and social experience categorically and schematically, and it made all kinds of productive and hypothetical inferences and chains of inferences about its experience as well—all with a modicum of cognitive self-monitoring. And so, the crucial point is that well before the emergence of uniquely human sociality, much less culture, language, and institutions, the foundations for human thinking were securely in place in humans' last common ancestor with other apes.

Individual intentionality is what is needed for creatures whose social interactions are mainly competitive, that is, creatures that act on their own or, at
most, join in with others to choose sides when there is a good fight going
on. In virtually all theoretical accounts, great apes' skills of social cognition
evolved mainly for competing with others in the social group: being better or
quicker than groupmates at anticipating what potential competitors might do,
based on a kind of Machiavellian intelligence (Whiten and Byrne, 1988). And
indeed a number of recent studies have found that great apes utilize their most
sophisticated skills of social cognition in contexts involving competition or
exploitation of others as opposed to contexts involving cooperation or communication with others (e.g., Hare and Tomasello, 2004; see Hare, 2001).
Great apes are all about cognition for competition.

Human beings, in contrast, are all about (or mostly about) cooperation.
Human social life is much more cooperatively organized than that of other
primates, and so, in the current hypothesis, it was these more complex forms
of cooperative sociality that acted as the selective pressures that transformed
great ape individual intentionality and thinking into human shared intentionality and thinking. Our task now is thus to provide a plausible evolutionary narrative that can take us from humans' great ape ancestors all the way to
modern humans. The *shared intentionality hypothesis* is that this story comprises
a two-step evolutionary sequence: joint intentionality followed by collective
intentionality. At both of these transitions the overall process was, at a very
general level, the same: a change of ecology led to some new forms of collaboration, which required for their coordination some new forms of cooperative
communication, and then together these created the possibility that, during ontogeny, individuals could construct through their social interactions
with others some new forms of cognitive representation, inference, and self-
monitoring for use in their thinking.

# 3

# Joint Intentionality

> Conceptual contents are essentially expressively perspectival.
>
> —ROBERT BRANDOM, *MAKING IT EXPLICIT*

In their magisterial survey of life on planet earth, Maynard-Smith and Szathmary (1995) identified eight major transitions in the evolution of complexity of living things, for example, the emergence of chromosomes, the emergence of multicellular organisms, and the emergence of sexual reproduction. Astoundingly, in each case the transition was characterized by the same two fundamental processes. First, in each case there emerged some new form of cooperation with interdependence: "Entities that were capable of independent replication before the transition can replicate only as part of a larger whole after it" (p. 6). Second, in each case this new form of cooperation was made possible by a concomitant new form of communication: "change in the method of information transmission" (p. 6).

The most recent major transition, in this account, was the emergence of human cooperative societies (cultures) structured by linguistic communication. Our ultimate goal is to give an account of this emergence, with a specific focus on the new forms of thinking that it engendered. But we cannot go directly from competitive great ape societies to cooperative human cultures in one giant leap. The problem is that there are thousands of human cultures, and each of them has conventionalized, normativized, and institutionalized a particular set of cultural and communicative practices. But anything may be conventionalized, normativized, or institutionalized; these processes are totally blind to content. And so to get to cooperatively organized human cultures, there must have been in existence in all human populations already—as raw material for these group-level processes of cultural creation—many and varied kinds of cooperative social interactions of a type not possessed by other great apes. Assuming again that great apes are representative of humans' last common ancestor with other primates, then, it would seem that we need an intermediate step in our natural history. We need some early humans who were not yet living in cultures and using conventional languages, but who were nevertheless much more cooperatively inclined than the last common ancestor.

And so we will posit in this chapter, as an initial step, some early humans who created new forms of social coordination, perhaps in the context of collaborative foraging. Early humans' new form of collaborative activity was unique among primates because it was structured by joint goals and joint attention into a kind of second-personal *joint intentionality* of the moment, a "we" intentionality with a particular other, within which each participant had an individual role and an individual perspective. Early humans' new form of cooperative communication—the natural gestures of pointing and pantomiming—enabled them to coordinate their roles and perspectives on external situations with a collaborative partner toward various kinds of joint objectives. The result was that these early humans "cooperativized" great ape individual intentionality into human joint intentionality involving new forms of cognitive representation (perspectival, symbolic), inference (socially recursive), and self-monitoring (regulating one's actions from the perspective of a cooperative partner), which, when put to use in solving concrete problems of social coordination, constituted a radically new form of thinking.

So let us look, first, at the new form of collaboration that emerged with early humans, then at the new form of cooperative communication that early humans used to coordinate their collaborative activities, and then at the resulting new form of thinking that all of this collaborating and communicating required as substrate.

## A New Form of Collaboration

Cooperation by itself does not create complex cognitive skills—witness the complex cooperation of the cognitively simple eusocial insects and the cooperative child care and food sharing of the not-so-cognitively-complex New World monkeys, marmosets and tamarins. The case of humans is unique, from a cognitive point of view, because the common ancestor to humans and other great apes had already evolved highly sophisticated skills of social cognition and social manipulation for purposes of competition (as well as highly sophisticated skills of physical cognition for purposes of manipulating causality in the context of tool use)—as documented in chapter 2.

Then, out of the elements of these sophisticated processes of individual intentionality built for competition—understanding how the particular goals and perceptions of others generate particular actions—humans evolved, in addition, even more sophisticated processes of *joint* intentionality, involving *joint* goals and *joint* attention, built for social coordination. And social coordination creates unique challenges for cognition and thinking. Whereas the social dilemmas of game theory (e.g., prisoner's dilemma) occur when interactants' goals and preferences mostly conflict, coordination dilemmas occur when individuals' goals and preferences mostly align. The challenge in these cases is not to resolve some conflict, but rather to find a way, perhaps by thinking, to coordinate with a social partner to a common goal.

### The Cooperative Turn

Chimpanzees and other great apes live in highly competitive societies in which individuals vie with others for valued resources all day every day, and, as argued above, this is what shapes their cognition most profoundly. But chimpanzees and other apes also engage regularly in a number of important activities that are cooperative in a very general sense. For example, chimpanzees travel together and forage in small groups, "allies" support one another in fights within the group, and males engage in group defense against outsiders and predators (Muller and Mitani, 2005). These group behaviors for traveling, fighting, and defending the group are also common in many other mammalian species.

To illustrate the difference with human cooperation, let us focus on foraging, clearly one of the fundamental activities of all primates. The typical scene for chimpanzees, for instance, is that a small traveling party comes upon a fruit tree. Each individual then scrambles up on its own, finds a good place to procure some fruit on its own, grabs one or several pieces on its own, and then separates from the others by a few meters to eat. In one recent experiment, when given a choice of acquiring food cooperatively or alone, chimpanzees preferred to acquire it alone (Bullinger et al., 2011a). In another recent experiment, when given a choice of eating together with a groupmate or eating alone, both chimpanzees and bonobos preferred to eat alone (Bullinger et al., 2013). If there is ever a conflict over a piece of food, the dominant individual (depending, ultimately, on fighting ability) gets it. In general, the acquisition of food via individual scrambling and contests of dominance characterizes virtually all of the foraging activities of the four great ape species.

The main exception to this general great ape pattern is chimpanzees' group hunting of monkeys—systematically observed only in chimpanzees, and only in some groups (Boesch and Boesch, 1989; Watts and Mitani 2002). What happens prototypically is that a small party of male chimpanzees spies a red colobus monkey somewhat separated from its group, which they then proceed to surround and capture. Normally, one individual begins the chase, and others scramble to the monkey's possible escape routes, including the ground. One individual actually captures the monkey, and he ends up getting the most and best meat. But because he cannot dominate the carcass on his own, all participants (and many bystanders) usually get at least some meat, depending on their dominance and the vigor with which they beg and harass the captor (Gilby, 2006).

The social and cognitive processes involved in chimpanzee group hunting could potentially be complex, but they could also be fairly simple. The "rich" reading is a human-like reading, namely, that chimpanzees have the joint goal of capturing the monkey together and that they coordinate their individual roles in doing so (Boesch, 2005). But more likely, in our opinion, is a "leaner" interpretation (Tomasello et al., 2005). In this interpretation, each individual is attempting to capture the monkey on its own (since captors get the most meat), and they take into account the behavior, and perhaps intentions, of the other chimpanzees as these affect their chances of capture. Adding some complexity, individuals prefer that one of the other hunters capture the monkey (in which case they will get a small amount of meat through begging and harassing) to the possibility of the monkey escaping totally (in which case they get no meat). In this view, chimpanzees in a group hunt are engaged in a kind of co-action in which each individual is pursuing his own individual goal of capturing the monkey (what Tuomela, 2007 calls "group behavior in I-mode"). In general, it is not clear that chimpanzees' group hunting of monkeys is so different cognitively from the group hunting of other social mammals, such as lions and wolves.

In stark contrast, human foraging is collaborative in much more fundamental ways. In modern forager societies, individuals produce the vast majority of their daily sustenance collaboratively with others, either immediately through collaborative efforts or via procurers who bring the food back to some central location for sharing (Hill and Hurtado, 1996; Hill, 2002; Alvard,

2012).[1] Human foragers also collaborate in many other domains of activity in ways that great apes do not. Tomasello (2011) systematically compares the social structures of great ape and human forager societies and concludes that in every domain, whereas apes behave mostly individualistically, humans behave mostly cooperatively. For example, humans but not apes engage in cooperative childcare in which all adults do all kinds of things to support developing children (so-called cooperative breeding; Hrdy, 2009). Humans but not apes engage in cooperative communication in which they provide one another with information that they judge to be useful for the recipient. Humans but not apes actively teach one another things helpfully, again for the benefit of the recipient. Humans but not apes make group decisions about group-relevant matters. And humans but not apes create and maintain all kinds of formal social structures such as social norms and institutions and even conventional languages (using agreed-upon means of expression). In all, cooperation is simply a defining feature of human societies in a way that it is not for the societies of the other great apes.[2]

Exactly when and how this cooperative turn took place in human evolution are not critical for current purposes. But for whatever it is worth, Tomasello et al. (2012) hypothesize that it began happening in an initial, preparatory step soon after the emergence of the genus *Homo*, around 2 million years ago. During this period there was a great expansion of terrestrial monkeys, like baboons, that might have outcompeted humans for their normal fruits and other vegetation. Humans then needed a new foraging niche. A beginning might have been scavenging meat, which would probably have required a kind of coalition of individuals to frighten off the animals that made the initial kill. But at some point there began more active collaborative hunting of large game and gathering of plant foods, typically in a mutualistic stag hunt–type situation in which both individuals could expect to benefit from the collaboration—if they could somehow manage to coordinate their efforts. This is the collaborative creature we are imagining here, and for the most clarity we may focus on its culmination in hominins of about 400,000 years ago: the common ancestor to Neanderthals and modern humans, the ever mysterious *Homo heidelbergensis*. Paleoanthropological evidence suggests that this was the first hominin to engage systematically in the collaborative hunting of large game, using weapons that almost certainly would not enable a single individual to be successful on its own, and sometimes bringing prey back to a home base (Stiner et al., 2009). This is also a time when brain size and population size were both expanding rapidly (Gowlett et al., 2012). We may hypothesize that these collaborative foragers lived as more or less loose bands comprising a kind of pool of potential collaborators.

But more important than when is how. In the hypothesis of Tomasello et al. (2012), obligate collaborative foraging became an evolutionarily stable strategy for early humans because of two interrelated processes: interdependence and social selection. The first and most basic point is that humans began a lifestyle in which individuals could not procure their daily sustenance alone but instead were interdependent with others in their foraging activities— which meant that individuals needed to develop the skills and motivations to forage collaboratively or else starve. There was thus direct and immediate selective pressure for skills and motivations for joint collaborative activity (joint intentionality). The second point is that as a natural outcome of this interdependence, individuals began to make evaluative judgments about others as potential collaborative partners: they began to be socially selective, since choosing a poor partner meant less food. Cheaters and laggards were thus selected against, and bullies lost their power to bully. Importantly, this now meant that early human individuals had to worry, in a way that other great apes do not, both about evaluating others and about how others were evaluating them as potential collaborative partners (i.e., a concern for self-image).

The situation these early humans faced is perhaps best modeled by the stag hunt scenario from game theory (Skyrms, 2004). Two individuals have easy access to low-payoff "hares" (e.g., low-calorie vegetation), and then there appears on the horizon a high-payoff but difficult-to-obtain "stag" (e.g., large game) that can be acquired only if individuals abandon their hares and collaborate. Their motivations thus align, because it is in both their interests to work together. The dilemma is thus purely cognitive: since collaboration is mandatory and I am risking my hare, I want to go for the stag only if you do, too. But you only want to go for the stag if I do, too. How do we coordinate this potential standoff? There are some cognitively simple ways out of the dilemma (see Bullinger et al., 2011b, for the leader-follower strategy that chimpanzees use), but they always involve one individual incurring disproportionate risk, and so they are unstable in certain circumstances. For example, if there were very few hares, so that each was highly valued, and hunting stags was only rarely successful, then the cost/benefit analysis would require that each individual attempt to make certain that their potential partner was also going for the stag *before* they relinquished their hare.

In the original analyses of Schelling (1960) and Lewis (1969), coordinating in this way required some kind of mutual knowledge or recursive mind reading: for me to go, I have to expect you to expect me to expect you. . . . For both Schelling and Lewis, this process, while remarkable, did not cause alarm. Later commentators problematized the analysis, pointing out that an infinite back-and-forth of us thinking about one another's thinking could not actually be happening, or no decision could ever be made. Clark (1996) proposed, as a more realistic account, that humans simply recognize the "common ground" they have with others (e.g., we both know that we both want to go for the stag) and that this is sufficient for making joint decisions toward joint goals. Tomasello (2008) suggested that something like common ground is how people actually operate, but when perturbations occur they often explain them by reasoning that "he thinks that I think he thinks . . ." (typically only a few iterations deep), suggesting an underlying recursive structure. Our position is thus that human individuals are attuned to the common ground they share with others, and this does not always involve recursive mind reading, but still, if necessary, they may decompose their common ground a few recursive layers deep to ask such things as what he thinks I think about his thinking.

In any case, we may imagine that individuals who were attuned to their common ground with others, and who could engage in some level of recursive mind reading, had a huge advantage in strategically deciding when to stay with their hare and when to join with others to go for the more profitable stag. And those who could develop more sophisticated forms of cooperative communication would have had an even larger advantage. And so, the first step in our natural history of uniquely human thinking is the cognitive mechanisms of joint intentionality that evolved to coordinate humans' earliest species-unique forms of small-scale collaboration and, later, cooperative communication.

## Joint Goals and Individual Roles

We may characterize the formation of a joint goal (or joint intention) in more detail as follows (see Bratman, 1992). For you and me to form a joint goal (or joint intention) to pursue a stag together, (1) I must have the goal to capture the stag together with you; (2) you must have the goal to capture the stag together with me; and, critically, (3) we must have mutual knowledge, or common ground, that we both know each other's goal.

It is important here that each of our goals is not just to capture the stag but, rather, to capture it together with the other. Each of us wanting separately to capture the stag (even if this was mutual knowledge; see Searle, 1995) would constitute two individuals hunting in parallel, not jointly. It is also important that we have mutual knowledge of one another's goal, that is, that our respective goals are part of our common conceptual ground. Each of us may want to capture the stag together with the other, but if neither of us knows that this is the case, we very likely will not succeed in coordinating (for all of the reasons outlined by Lewis and Schelling, among others). Thus, joint intentionality is operative both in the action content of each of our goals or intentions—that we act together—and in our mutual knowledge, or common ground, that we both know that we both intend this.

Young children begin engaging with others in ways that suggest some form of joint goal from around fourteen to eighteen months of age, when they are still mostly prelinguistic. Thus, Warneken et al. (2006, 2007) had infants of this age engage in a joint activity with an adult, such as obtaining a toy by each operating one side of an apparatus. Then, the adult simply stopped playing her role for no reason. The children were not happy about this and did various things to attempt to reengage their partner. (They did not do this if her stopping was for a good reason; e.g., she had to attend to something else [Warneken et al., 2012].) Interesting, when this same situation was arranged for human-raised chimpanzees, they simply ignored the recalcitrant partner and tried to find ways to achieve the goal on their own. Although infants' reengagement attempts do not suggest necessarily that they have a fully adult-like joint goal in common ground with their partner, at the very least they reflect an expectation that, barring obstacles, my partner in this joint activity is committed enough to reengage after a stoppage—an expectation that, apparently, chimpanzees in similar activities do not have.

By the time they are three years of age, children provide much more convincing evidence for joint goals because they themselves display commitment to the joint activity in the face of distractions and temptations. For example, Hamann et al. (2012) had pairs of three-year-old children work together to bring rewards to the top of a step-like structure. The problem was that for one child the reward, surprisingly, became available midway through. Nevertheless, when this happened, the lucky child delayed consumption of her own reward and persevered until the other got hers (i.e., more than they helped the partner in a similar situation in which they were acting individually, without collaboration). Such commitment to the partner suggests that the children constructed a joint goal at the beginning that "we" get the prizes together, and they made whatever adjustments were necessary to realize that joint goal. Again, great apes do not behave in this same way. In a similar experiment with chimpanzees, Greenberg et al. (2010) found no signs of a human-like commitment to follow through on the joint action until both partners received their reward. (And Hamann et al. [2011] found that at the end of the collaborative activity, three-year-old children, but not chimpanzees, were committed to dividing the spoils equally among participants as well.)

Importantly, when children of this same age have it in their common ground with a collaborative partner that each is counting on the other to come through (we are interdependent), they both feel obligated to the other (see Gilbert, 1989, 1990). Thus, Gräfenhain et al. (2009) had preschoolers explicitly agree to play a game with one adult, and then another adult attempted to lure them away to a more exciting game. Although two-year-old children mostly just bolted to the new game straightaway, from three years of age children paused before departing and "took leave," either verbally or by handing the adult the tool they had been using together. The children seemed to recognize that joint goals involve joint commitments, the breaking of which requires some kind of acknowledgment or even apology. No study of this type has ever been done with chimpanzees, but there are no published reports of one chimpanzee taking leave from, making excuses to, or apologizing to another for breaking a joint commitment.

In addition to joint goals, collaborative activities also demand a division of labor and so individual roles. Bratman (1992) specifies that in joint cooperative activities individuals must "mesh" their subplans together toward the joint goal, and even help one another in their individual roles as necessary. In the Hamann et al. (2012) study cited above, young children stopped to help their partner as needed. This demonstrates that the partners are attending to one another and their respective subgoals, and perhaps even attending to the partner attending to them, and so forth. Indeed, other studies have found that young children, but not chimpanzees, learn important new things about the partner's role as they are collaborating. For example, Carpenter et al. (2005) found that after young children played one role in a collaboration, they could quickly switch to the other, whereas chimpanzees could not do this (Tomasello and Carpenter, 2005). Most important, Fletcher et al. (2012) found that three-year-olds who had first participated in a collaboration playing role

A then knew much better how to play role B than if they had not previously collaborated, whereas this was not true of chimpanzees.

Young children are thus beginning to understand that the roles in a collaborative activity are in most cases interchangeable among individuals, which suggests a "bird's eye view" of the collaboration in which the various roles, including one's own, are all conceptualized in the same representational format (see Hobson, 2004). This species-unique understanding may support an especially deep appreciation of self-other equivalence, as individuals imagine different subjects/agents engaging in similar or complementary activities *simultaneously* in the same collaborative activity. As suggested in our discussion of great ape thinking, the understanding of self-other equivalence is a key component enabling various kinds of combinatorial flexibility in thinking. (It also sets the stage for a full-blooded appreciation of agent neutrality encompassing not just self and other but all possible agents, which is a key feature of cultural norms and institutions, and "objectivity" more generally, as we shall see in chapter 4.)

Preschool children are not good models for the early humans we are picturing here because they are modern humans and they are bathed in culture and language from the beginning. But from soon after their first birthdays, and continuing up to their third birthdays, they come to engage with others in collaborative activities that have a species-unique structure and that do not, in any obvious way, depend on cultural conventions or language. These young children coordinate a joint goal, commit themselves to that joint goal until all get their reward, expect others to be similarly committed to the joint goal, divide the common spoils of a collaboration equally, take leave when breaking a commitment, understand their own and the partner's role in the joint activity, and even help the partner in her role when necessary. When tested in highly similar circumstances, humans' nearest primate relatives, great apes, do not show any of these capacities for collaborative activities underlain by joint intentionality. Importantly, young children also seem to have a species-unique motivation for collaboration, as shown in recent studies in which children and chimpanzees had to choose between pulling in a certain amount of food collaboratively with a conspecific or pulling in that same amount of food (or more or less) in a solo activity. Children very much preferred the collaborative option, whereas chimpanzees went wherever there was most food regardless of opportunities for collaboration (Rekers et al., 2011; Bullinger et al., 2011a).

BOX I.  Relational Thinking

Penn et al. (2008) have proposed that what makes human cognition different from that of other primates is thinking in terms of relations, especially higher-order relations. To support their claim, they review evidence from many different domains of cognition: judgments of relational similarity, judgments of same-difference, analogy, transitive inference, hierarchical relations, and so forth.

Their evaluation of the literature is decidedly one-sided, as they dismiss findings suggesting that nonhuman primates have some skills of this type. For example, nonhuman primates clearly understand some relations (consistently choosing the larger of two objects, for example, despite absolute size), and some individuals make same-difference judgments again based on relational not absolute characteristics (Thompson et al., 1997). Some chimpanzees also do something like analogical inference in using a scale model (Kuhlmeier et al., 1999), and many primates make transitive inferences (see Tomasello and Call, 1997, for a review).

But at the same time, it is true that humans are particularly skilled at relational thinking (Gentner, 2003). One hypothesis that might explain the data is that there are actually two kinds of relational thinking. One concerns the concrete physical world of space and quantities, in which we may compare various characteristics or magnitudes such as bigger-smaller, brighter-darker, fewer-greater, higher-lower, and even same-different. Nonhuman primates have some skills with these kinds of physical relations and relational magnitudes. What they may not comprehend at all—though there are few direct tests—is a second type of relation. Specifically, they may not comprehend functional categories of things defined by their role in some larger activity. Humans are exceptional in creating categories such as pet, husband, pedestrian, referee, customer, guest, tenant, and so forth, what Markman and Stillwell (2001) call "role-based categories." They are relational not in the sense of comparing two physical entities but, rather, in assessing the relation between an entity and some larger event or process in which it plays a role.

The obvious hypothesis here is that this second type of relational thinking comes from humans' unique understanding of collaborative activities with joint goals and individual roles (perhaps later generalized to all kinds of social activities even if they are not collaborative per se). As humans constructed these kinds of activities, they were creating more or less abstract "slots" or roles that anyone could play. These abstract slots formed role-based categories, such as things that one uses to kill game (viz., weapons; Barsalou, 1983), as well as more abstract narrative categories such as protagonist, victim, avenger, and so on. A further speculation might be that these abstract slots at some point enabled humans to even put relational material in the slots; for example, a married couple can play a role in a cultural activity. This would be the basis for the kinds of higher-order relational thinking that Penn et al. (2008) emphasize as especially important in differentiating human thinking.

In any case, the proposal here is that, at the very least, constructing the kinds of dual-level cognitive models needed to support collaborative activities enhanced, if not enabled, human engagement in much broader and more flexible relational thinking involving roles in larger social realities, and possibly in higher-order relational thinking as well.

The main point for now is that early humans seem to have created a new cognitive model. Collaborating toward a joint goal created a new kind of social engagement, a joint intentionality in which "we" are hunting antelopes together (or whatever), with each partner playing her own interdependent role. This dual-level structure of simultaneous sharedness and individuality—a joint goal but with individual roles—is a uniquely human form of second-personal joint engagement requiring species-unique cognitive skills and motivational propensities. It also has a number of perhaps surprising ramifications for many different aspects of human cognition that go beyond our primary focus here (see box 1 for one example).

### Joint Attention and Individual Perspectives

Organisms attend to situations that are relevant to their goals. And so, when two humans act together jointly, they naturally attend together jointly to situations that are relevant to their joint goals. Said another way, as humans coordinate their actions, they also, to do this effectively, coordinate their attention. Underlying this coordination is, once again, some notion of common ground, in which each individual—at least potentially—can attend to his partner's attention, his partner's attention to his attention, and so forth (Tomasello, 1995). Joint actions, joint goals, and joint attention are thus of a piece, and so they must have coevolved together.

The current proposal is that the phylogenetic origins of the ability to participate with others in joint attention—the first and most concrete way in which young children create common conceptual ground and so shared realities with others—lie in collaborative activities. This is what Tomasello (2008) calls the "top-down" version of joint attention because it is directed by joint goals. (The alternative is bottom-up joint attention, such as when a loud noise attracts both of our attention, and we both know it must have attracted the other's attention as well.) Ontogenetically, young children begin to structure their joint actions with others via joint visual attention at around nine to twelve months of age, often called joint attentional activities. These are such activities as giving and taking objects, rolling a ball back and forth, building a block tower together, putting away toys together, and "reading" books together. Despite specific attempts to identify and solicit such joint attentional activities with human-raised chimpanzees, Tomasello and Carpenter (2005) were unable to find any (nor are there any other reliable reports of joint attention in nonhuman primates).

Just as each partner in a joint collaborative activity has her own individual role, each partner in joint attentional engagement has her own individual perspective—and knows that the other has her own individual perspective as well. The crucial point, which will be foundational for all that follows, is that *the notion of perspective assumes a single target of joint attention on which we have differing perspectives* (Moll and Tomasello, 2007, in press). If you are looking out one window of the house and I am looking out another in the opposite direction, we do not have different perspectives—we are just seeing completely different things. We can thus operate with the notion of individually distinct perspectives only if (1) we are both considering "the same" thing, and (2) we both know the other is attending to it differently. If I see something in one way, and then round the corner to see it in another, this does not give me two perspectives on the same thing, because I do not have multiple perspectives available to me simultaneously for comparison. But when two people are attending to the same thing simultaneously—and it is in their common ground that they are both doing so—then "space is created" (to use

Davidson's [2001] metaphor) for an understanding of different perspectives to arise.[3]

Young children begin showing an appreciation that others have perspectives that differ from their own from soon after their first birthdays, in conjunction with their earliest joint attentional activities. For example, in one experiment an adult and child played together jointly with three different objects for a short time each (Tomasello and Haberl, 2003). Then, while the adult was out of the room, the child and a research assistant played with a fourth object. After that the adult returned, looked at an array containing all four objects, and exclaimed excitedly "Wow! Cool! Look at that!" Under the assumption that people only get excited about new things, not old things, children as young as twelve months of age identified which of the objects was the new one causing the adult's excitement—even though they were all equally old for her. The new one is the one we have not attended to together before.

This is what some researchers have called level 1 perspective taking, because it concerns only whether the other person does or does not see something, not *how* she sees it. In level 2 perspective taking, children understand that someone sees the same thing differently than they do. For example, Moll et al. (2013) found that three-year-old children understood which object an adult intended to indicate by calling it "blue," even though the object did not appear blue to the child—only to the adult due to her looking through a color filter. Children could thus take the perspective of the other person when it differed from their own. However, these same children could not answer correctly when asked if they and the adult saw the object differently *at the same time*. Indeed, children struggle with several versions of such simultaneously conflicting perspectives on a jointly attended-to situation until they are four or five years of age (Moll and Tomasello, in press). Thus, children before four or five years have difficulty with dual naming tasks ("it" is simultaneously a *horse* and a *pony*), appearance-reality tasks ("it" is simultaneously a rock and a sponge; Moll and Tomasello, 2012), and false belief tasks ("it" is in the cabinet or in the box). Resolving the conflict between perspectives that confront one another on a jointly attended-to entity—especially when both purport to depict "reality"—takes some additional skills for dealing with an objective reality and how different perspectives relate to it, which again will await the next step in our evolutionary story (and see note 6).

And so we have come to a tipping point. Based on their capacity to coordinate actions and attention with others toward joint goals, early humans came to understand that different individuals can have different perspectives on one and the same situation or entity. In contrast, great apes (including the last common ancestor with humans) do not coordinate their actions and attention with others in this same way, and so they do not understand the notion of simultaneously different perspectives on the same situation or entity at all. We thus encounter once again the dual-level structure of simultaneous jointness and individuality. Just as collaborative activities have the dual-level structure of joint goal and individual roles, joint attentional activities have the dual-level structure of joint attention and individual perspectives. Joint attention thus begin the process by which human beings construct an inter-subjective world with others—shared but with differing perspectives—which will also be fundamental to human cooperative communication. We may thus posit that joint attention in joint collaborative activity, as manifest even in very young children, was the most basic form of socially shared cognition in human evolution—characteristic already of early humans—and that this primal version of socially shared cognition spawned an equally primal version of perspectively constructed cognitive representations.

### Social Self-Monitoring

Early humans living as obligate collaborative foragers would have become more deeply social in still another way. Although skills of joint intentionality are necessary for human-like collaborative foraging, they are not sufficient. One also has to find a good partner. This may not always be overly difficult, as even chimpanzees, after some experience, learn which partners are good (i.e., lead to success) and which are not (Melis et al., 2006b). But in addition, in situations in which there is meaningful partner choice, one must be—or at least appear to be—a good collaborative partner oneself. To be an attractive partner for others, and so not be excluded from collaborative opportunities, one must not only have good collaborative skills, but also do one's share of the work, help one's partner when necessary, share the spoils at the end of the collaboration, and so forth.

And so early humans had to develop a concern for how other individuals in their group were evaluating them as potential collaborative partners, and then regulate their actions so as to affect these external social judgments in positive ways—what we may call social self-monitoring. Other great apes do not appear to engage in such social self-monitoring. Thus, when Engelmann et al. (2012) gave apes the opportunity to either share or steal food from a groupmate, their behavior was totally unaffected by the presence or absence of other group members observing the process. In contrast, in the same situations, young human children shared more with others and stole less from others when another child was watching.

Motivationally, a concern for social evaluation derives from the interdependence of collaborative partners: my survival depends on how you judge me. Cognitively, a concern for social evaluation involves still another form of recursive thinking: I am concerned about how you are thinking about my intentional states. Social self-monitoring is thus the first step in humans' tendency to regulate their behavior not just by its instrumental success, as apes do in their goal-directed activities, but also by the anticipated social evaluations of important others. Because these concerns are about the evaluations of specific other individuals, we may think of them as second-personal phenomena. They thus represent an initial sense of social normativity—a concern for what others think I should and should not be doing and thinking—and so a first step toward the kind of normative self-governance, so as to fit in with group expectations, that will characterize modern humans in the next step of our story (see chapter 4).

### Summary: Second-Personal Social Engagement

Great apes have a multitude of social-cognitive skills for understanding the intentional actions of others, but they do not engage with them in any form of joint intentionality. Thus, great apes understand that others have goals, and they sometimes even help others attain their goals (Warneken and Tomasello, 2009), but they do not collaborate with others by means of a joint goal. Similarly, great apes understand that others see things and so can follow the gaze direction of others to see what they see (Call and Tomasello, 2005), but they do not engage with others in joint attention. And great apes make individual decisions that they self-monitor, but they do not make joint decisions with others or monitor themselves via the social evaluations of others. What emerges for the first time with early humans, in the current account, is a "we" intentionality in which two individuals engage with the intentional states of one another both jointly and recursively.

This new form of joint engagement is second-personal: an engagement of "I" and "you." Second-personal engagement has two minimal characteristics: (1) the individual is directly participating in, not observing from outside, the social interaction; and (2) the interaction is with a specific other individual with whom there is a dyadic relationship, not with something more general like a group (if there are multiple persons present, there are many dyadic relationships but no sense of group). There is less consensus about other possible features of second-personal engagement, but Darwall (2006) proposes, in addition, that (3) the essence of this kind of engagement is "mutual recognition" in which each partner gives the other, and expects from the other, a certain amount of respect as an equal individual—a fundamentally cooperative attitude among partners.

And so, the evolutionary proposal is that early humans—perhaps *Homo heidelbergensis* some 400,000 years ago—evolved skills and motivations for joint intentionality that transformed great apes' parallel group activities (e.g., you and I are each chasing the monkey in parallel) into truly joint collaborative activities (e.g., we are chasing the monkey together, each with our own role). And they also transformed great apes' parallel looking behavior (e.g., you and I are each looking at the banana) into true joint attention (e.g., we are looking at the banana together, each with our own perspective). But early humans were not doing this in the manner of contemporary human beings, that is, as manifest in relatively permanent cultural conventions and institutions. Rather, their earliest collaborative activities were ad hoc collaborations for particular goals on a particular occasion with a particular person, with their joint attention similarly structured in this second-personal way. There would thus be second-personal joint engagement with the partner, but when the collaboration was over the "we" intentionality would be over as well.

The cognitive model schematized from repeated experiences of this kind was thus a dual-level structure of simultaneous sharing and individuality while in direct social interaction with other particular partners, underlain and supported by some kind of common ground and recursive mind reading (see Figure 3.1). The cognitive model of this second-personal, dual-level social engagement laid the foundation for almost everything that was uniquely human. It provided the joint intentionality infrastructure for uniquely human forms of cooperative communication involving intentions and inferences about perspectives—as we shall see in the section that follows—and, ultimately, it provided the foundation for the cultural conventions, norms, and institutions

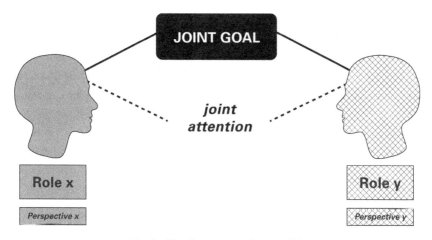

FIGURE 3.1 The dual-level structure of joint collaborative activity

that brought the human species into the modern human world—as we shall
see in the next chapter.

## A New Form of Cooperative Communication

Early humans coordinated their actions and attention based on common
ground. But coordinating in more complex ways—for example, in planning
our specific roles in a collaboration under various contingencies, or in planning
a series of joint actions—required a new type of cooperative communication.
The gestures and vocalizations of ancient great apes could not have done this
coordinating work. They could not have, first, because they were geared
totally toward self-serving ends, and this simply does not mesh with mutualistic collaboration toward a joint goal. They could not have, second, because
they were used exclusively in attempts to regulate the behavior of others
directly, and this does not mesh with the need to coordinate actions and
attention referentially on external situations and entities, as in collaborative
foraging for food.

Tomasello (2008) argued and presented evidence that the first forms of
uniquely human cooperative communication were the natural gestures of
pointing and pantomiming used to inform others helpfully of situations relevant to them. Pointing and pantomiming are human universals that even

people who share no conventional language can use to communicate effectively in contexts with at least some common ground. But to do this requires an extremely rich and deep set of interpersonal intentions and inferences in the context of this common ground. If I point for you in the direction of a tree, or pantomime for you a tree, without any common ground, you have nothing to guide your inferences about what I am intending to communicate to you or why. Pointing and pantomiming thus created for early humans new problems of social coordination—not just in coordinating actions with others but also in coordinating intentional states—and solving these new coordination problems required new ways of thinking.

### A New Motive for Communicating

In joint collaborative activities in which the partners are interdependent, it is in the interest of each partner to help the other play her role. This is the basis for a new motive in human communication, not available to other apes (but see Crockford et al., 2011, for one possible exception), namely, the motive to help the other by informing her of situations relevant to her. The emergence of this motive was aided by the fact that in the context of a joint collaborative activity, directive communication and informative communication are not clearly distinct—because the partners' individual motives are so closely intertwined. Thus, if we are gathering honey together and you are struggling with your role, I can point to a stick, which I intend as a directive to you to use it, or, alternatively, I can point to the stick intending only to inform you of its presence—because I know that if you see it you will most likely want to use it. When we are working together toward a joint goal, both of these work because our interests are so closely aligned.

The evolutionary proposal is thus that early humans' first acts of cooperative communication were pointing gestures in joint collaborative activities, and these were underlain by a communicative motive not yet differentiated between requestive and informative. But at some point early humans began to understand their interdependence with others not just while the collaboration was ongoing but also more generally: if my best partner is hungry tonight, I should help her so that she will be in good shape for tomorrow's foraging. And outside of collaborative activities, the difference between me requesting help from you, for my benefit, and me informing you of things helpfully, for your benefit, becomes crystal clear. And so there arose for early humans two distinct motives for their deictic communication, requestive and informative, which everyone both comprehended and produced.

When great apes work together in experiments, there is an almost total absence of intentional communication of any kind (e.g., Melis et al., 2009; Hirata, 2007; Povinelli and O'Neill, 2000). When apes communicate with one another in other contexts, it is always directive (Call and Tomasello, 2007; Bullinger et al., 2011c). In stark contrast, from as soon as they can collaborate meaningfully with others, at around fourteen to eighteen months of age, young children use the pointing gesture to coordinate their joint activity (e.g., Brownell and Carriger, 1990; Warneken et al., 2006, 2007)—with, again, a telling ambiguity about whether their motive is requestive or informative. But also, outside of collaborative activities, even twelve-month-old infants sometimes point simply to inform others of such things as the location of a sought-for object. For example, Liszkowski et al. (2006, 2008) placed twelve-month-olds in various situations in which they observed an adult misplace an object or lose track of it in some way, and then start searching. In these situations infants pointed to the sought-for object (more often than to distractor objects that were misplaced in the same way but were not needed by the adult), and in doing this they showed no signs of wanting the object for themselves (no whining, reaching, etc.). The infants simply wanted to help the adult by informing her of the location of the sought-for object.

The emergence of the informative communicative motive, alongside the general great ape directive motive, had three important consequences for the evolution of uniquely human thinking. First, the informative motive led communicators to make a commitment to informing others of things honestly and accurately, that is, truthfully. Initially during collaborative activities, but then more generally (as humans' interdependence extended outside of collaborative activities), if individuals wanted to be seen as cooperative, they would commit themselves to always communicating with others honestly. Of course, you may still lie: you point to where you want me to search for my spear even though it is not really there, for some selfish motive. But lying only works if there is first a mutual assumption of cooperation and trust: you only lie because you know that I will trust your information as truthful and act accordingly. And so, while there is still some way to go to get to truth as an "objective" feature of linguistic utterances (see chapter 4), if we want to explain the origins of humans' commitment to characterize the world accurately independent of any selfish purpose, then being committed to informing others of things honestly, for *their* not *our* benefit, is the starting point. The notion of truth thus entered the human psyche not with the advent of individual intentionality and its focus on accuracy in information acquisition but, rather, with the advent of joint intentionality and its focus on communicating cooperatively with others.[4]

The second important consequence of this new cooperative way of communicating was that it created a new kind of inference, namely, a relevance inference. The recipient of a cooperative communicative act asks herself: given that we know together that he is trying to help me, why does he think that I will find the situation he is pointing out to me relevant to *my* concerns. Consider great apes. If a human points and looks at some food on the ground, apes will follow the pointing/looking to the food and so take it—no inferences required. But if food is hidden in one of two buckets (and the ape knows it is in only one of them) and a human then points to a bucket, apes are clueless (see Tomasello, 2006, for a review). Apes follow the human's pointing and looking to the bucket, but then they do not make the seemingly straightforward inference that the human is directing their attention there because he thinks it is somehow relevant to their current search for the food. They do not make this relevance inference because it does not occur to them that the human is trying to inform them helpfully—since ape communication is always directive—and this means that they are totally uninterested in why the human is pointing to one of the boring buckets. Importantly, it is not that apes cannot make inferences from human behavior at all. When a human first sets up with them a competitive situation and then reaches desperately toward one of the buckets, great apes know immediately that the food must be in that one (Hare and Tomasello, 2004). They make the competitive inference, "He wants in that bucket; therefore the food must be in there," but they do not make the cooperative inference, "He wants me to know that the food is in that bucket."

This pattern of behavior contrasts markedly with that of human infants. In the same situation, prelinguistic infants of only twelve months of age trust that the adult is pointing out to them something relevant to their current search—they comprehend the informative motive—and so they know immediately that the pointed-to bucket is the one containing the reward (Behne et al., 2005, 2012). The mutual assumption of cooperativeness in such situations is so natural for humans that they have developed a special set of signals—ostensive signals such as eye contact and addressing the other vocally—by means of which the communicator highlights for the recipient that he has some relevant information for her. Thus, as evolutionary example, suppose that while we are collaboratively foraging I point to berries on a bush for you, with eye contact and an excited vocalization. You look and see the bush but, at first, no berries. So you ask yourself: why does he think that this bush is relevant for me—and this makes you look harder for something that is indeed relevant—and thus you discover the berries. As communicator, I know that you, as recipient, are going to be engaging in this process if and only if you see me as directing your attention cooperatively, and so I want to make sure that you know that I am doing this. Therefore, I not only want you to know that there are berries here but also want you to know that I want you to know this—so that you will follow through the inferential process to its conclusion (Grice, 1957; Moore, in press). By addressing you ostensively, and based on our mutual expectation of cooperation, I am in effect saying, "You are going to want to know this"—and you do want to know it because you trust that I have your interests in mind.

The third and final consequence of this newly cooperative way of communicating was that there now emerged, at least in nascent form, a distinction between communicative force—as overtly expressed in requestive and informative intonations—and situational or propositional content as indicated by the pointing gesture. (NB: This means that by this time early humans would have had to control their vocal expressions of emotions voluntarily in a way that apes do not.) Early humans could now point toward berries in a bush, with one of two different motives, expressed intonationally: either an insistent requestive intonation, in the hopes that the recipient would fetch some berries for her, or a neutral intonation to just inform the recipient of the berries' location so that she might get some for herself. We thus now have a clear distinction between something like communicative force and communicative content: the communicative content is the presence of the berries, and the communicative force is either requestive or informative. All of this is implicit, of course, and so we still have some way to go to reach the conventionalized and so explicit distinction between communicative force and content that is so important in conventional linguistic communication (see chapter 4). But the breakthrough here is the relative independence of referential (situational, propositional) content from the communicator's motives or intentions for referring attention to it.

And so, early humans' joint collaborative activities created a new motivational infrastructure for their communication, a cooperative motivation to inform one another of things helpfully and honestly. This then motivated recipients to do significant inferential work to find out why the communicator thought that looking in a certain direction would be relevant for their concerns, which then motivated communicators to advertise when they had something relevant for a recipient. And the fact that there were now two different communicative motives possible—requestive and informative—meant that the situational (propositional) content of the communicative act was starting to be conceptualized as independent of the particular intentional states of the communicator.

### New Ways of Thinking for Communicating

The motive to be helpful and cooperative in communication meant that, from the cognitive point of view, communicators had to be able to determine which situations were relevant to a recipient on a particular occasion. Conversely, recipients had to be able to identify the intended situation and its relevance for them, in essence by determining which situation in the direction of his pointing gesture the communicator *thinks* is relevant or interesting for them on this occasion—and why. The basic problem is that what the communicator wishes to point out to the recipient—his communicative intention[5]—is a whole fact-like situation, for example, that there are bananas in the tree, or that there are no predators in the tree. But the act of pointing—the protruding finger—is the same in all cases. The puzzle is how does one point out for a recipient different situations in the same perceptual scene?

The key to this puzzle is that the participants in a communicative interaction mutually assume the relevance of the communicative act for the recipient (Sperber and Wilson, 1996), and this relevance is always in relation to something that is in our common ground (Tomasello, 2008). Independent of communication, a situation is relevant to you for your own individual reasons. But for me to direct your attention to that situation successfully in communication, you must know that I know it is relevant for you; indeed, we must know together in our common ground that it is relevant for you. The simplest situation is thus when we are in a collaborative activity with the immediate common ground created by our joint goal. For example, if we have been searching unsuccessfully for bananas all day, you will naturally assume that my pointing gesture toward the banana tree is intended to indicate for you the fact that there are bananas in that tree. On the other hand, if we spied the bananas together some minutes ago but there was a predator in the tree and so we waited, and now the predator seems to have left, you will naturally assume that I am indicating for you the fact that there is now no predator in the tree. Common ground and a mutual assumption of relevance—not possible for apes because they simply do not engage in this kind of cooperative communication— enable a meeting of minds in the direction of the protruding finger.

Following the analysis in chapter 2, relevant situations are those that present individuals with opportunities and/or obstacles for reaching their goals and maintaining their values. Thus, if during our search for fruit I point toward a distant banana tree, it would never occur to you that I might be pointing out the presence of the leaves, even if leaves is all that you see at the moment, since the presence of leaves is in no way relevant to what we are doing. Instead, you will continue looking until you see, for example, some bananas behind the leaves, whose presence is highly relevant to what we are doing. Another dimension of this process is that only "new" situations are communicatively relevant, since currently shared situations need not be pointed out. And so, in the example from above, after the predator left the banana tree I pointed to the banana tree with the intention to indicate the situation of the predator's absence, which you readily discerned. How could I intend and you infer predator absence when the presence of the bananas is also highly relevant? Because the presence of the bananas was already in our current common ground, and so me pointing out this situation to you would be superfluous. If I am going to be helpful, I must point out situations that are new for you, or else why bother. And so, in human cooperative communication, both communicators and recipients mutually assume in their common ground that communicators point out for recipients' situations that are both relevant and new.

Perhaps surprising, even young infants are skillful at keeping track of the common ground they have with specific other individuals and using that to determine relevance in both the comprehension and production of pointing gestures. For example, Liebal et al. (2009) had a one-year-old infant and an adult clean up together by picking up toys and putting them in a basket. At one point the adult stopped and pointed to a target toy, which the infant then cleaned up into the basket. However, when the infant and adult were cleaning up in exactly this same way, and a second adult who had not shared this context entered the room and pointed toward the target toy in exactly the same way, infants did not put the toy away into the basket; they mostly just handed it to him, presumably because the second adult had not shared the cleaning up game with them as common ground. Infants' interpretations thus did not depend on their own current egocentric activities and interests, which were the same in both cases, but rather on their shared experience with each of the pointing adults. (In another study, Liebal et al. [2010] found that infants of this same age also produced points differently depending on their common ground with the recipient.)

Infants in this same age range also use a mutual assumption of newness to determine what a pointing adult thinks is relevant for them. Thus, Moll et al. (2006) had eighteen-month-old infants play with an adult and a toy drum. If a new adult now entered the room and indicated the drum excitedly, the child assumed he was talking about the cool drum. But if the adult with whom the child had just been sharing enjoyment of the drum pointed to the drum excitedly in exactly the same manner, the child did not assume that she was excited about the drum: how could she be, since it is old news for us? Rather, children assumed that the adult's excitement must be due to something new about the drum that they had not previously noticed, and so they attended to some new aspect, for example, on the adult's side of the drum. In their production of pointing, infants also use this distinction between shared and new information. For example, when a fourteen-month-old infant wanted his mother to put his high chair up to the dining room table: on one occasion he pointed to the chair (because he and his mother had already shared attention to the empty space at the table), whereas on another occasion he pointed to the empty space at the table (because he and his mother had already shared attention to the chair) (Tomasello et al., 2007a). In both cases the infant wants the exact same thing—his chair placed at the table—but to communicate effectively he assumes that the object he and his mother are focused on is already part of their common ground, and so he points out the aspect of the situation that she may not have noticed, the new part.[6]

Engaging in cooperative (ostensive-inferential) communication of this type requires some new types of thinking. In effect, all three components of the thinking process—representation, inference, and self-monitoring—must become socialized.

With respect to representation, the key novelty is that both participants in the communicative interaction must represent one another's perspective on the situation and its elements. Thus, the communicator attempts to focus the recipient's attention on one of the many possible situations—fact-like representations—immanent in the current perceptual scene (e.g., there are bananas in the tree versus there is no predator in the tree). The communicative act thus perspectivizes the scene for the recipient. It also perspectivizes the elements. For example, if we are building a fire, me pointing out to you the presence of a log construes that log as firewood. But if we are tidying up the cave, me pointing out to you the presence of that very same log construes it as trash. In the object choice task, the communicator is not pointing to the bucket *qua* physical object or *qua* vessel for carrying water, but rather *qua* location: I am informing you that the reward is located *in there*. Cooperative pointing thus creates different conceptualizations or construals of things. These presage the ability of linguistic creatures to place one and the same entity under alternative different "descriptions" or "aspectual shapes," which is one of the hallmarks of human conceptual thinking; but it does this without the use of any conventional or symbolic vehicles with articulate semantic content.

With respect to inference, the key point is that the inferences used in cooperative communication are socially recursive. Thus, implicit in all of the foregoing is a kind of backing-and-forthing of individuals making inferences about the partner's intentions toward my intentional states. In the object choice task, for example, the recipient infers that the communicator *intends* that she *know* that the food is in that bucket—a socially recursive inference that great apes apparently do not make. This inference requires in all cases an abductive leap, something like: his pointing in the direction of that otherwise boring bucket would make sense (i.e., would be consistent with common ground, relevance, and newness) if it is the case that he *intends* that I *know* where the reward is. The communicator, for his part, is attempting to help the recipient to make that abductive leap appropriately. To do this, at least in many situations, the communicator must engage in some kind of simulation, or thinking, in which he imagines how pointing in a particular direction will lead the recipient to make a particular abductive inference: if I point in this direction, what inferences will he make about my intentions toward his intentional states? And then, when making his abductive inference, the recipient can potentially take into account the communicator's taking into account of what kind of inference she is likely to make about his communicative intentions. And so forth.

Finally, with respect to self-monitoring, the key is that being able to operate in this way communicatively requires individuals to self-monitor in a new way. As opposed to apes' cognitive self-monitoring, this new way was social. Specifically, as an individual was communicating with another, he was simultaneously imagining himself in the role of the recipient attempting to comprehend him (Mead, 1934). And so was born a new kind of self-monitoring in which communicators simulated the perspective of the recipient as a kind of check on whether the communicative act was well formulated and so was likely to be understood. This is not totally unlike the concern for self-image characteristic of early humans (noted earlier in the discussion of collaboration) in which individuals simulate how they are being judged by others for their cooperativeness—it is just that in this case what is being evaluated is comprehensibility. Importantly, both of these kinds of self-monitoring are "normative" in a second-personal way: the agent is evaluating his or her own behavior from the perspective of how other social agents will evaluate it. Of this process, Levinson (1995, p. 411) says, "There is an extraordinary shift in our thinking when we start to act intending that our actions should be coordinated with—then we have to design our actions so that they are self-evidently perspicuous."[7] This social self-monitoring for intelligibility in cooperative communication lays the foundation for modern human norms of social rationality, where social rationality means making communicative sense to one's partner.

These new processes of thinking involved in cooperative communication are well illustrated by two studies with young children. First, from the point of view of the communicator, is a study by Liszkowski et al. (2009) with twelve-month-old infants. In this study, an adult played a game with an infant in which the infant repeatedly needed a particular kind of object, always found in the same location on a plate. At some point, the infant needed one of those objects, but none was around. To get one, many infants alighted upon the strategy of pointing for the adult to the empty plate, that is, to the location where they both knew in common ground that those kind of objects usually are found. To perform this communicative act, the infant had to simulate the adult's process of comprehension: what abductive inference (about my intentions toward her intentional states) will she make if I point to the plate? That this is not just a simple association is suggested by the fact that chimpanzees, who are perfectly capable of associative learning, in this same setup made no attempts to direct the human's attention to the empty plate (even though they did make pointing attempts in other contexts in this same study). The children simulated the adult making inferences about their intentions toward her intentional states.

To illustrate the process even more dramatically, and from the point of view of comprehension, we may consider the phenomenon of "markedness." On some occasions, a communicator may mark (e.g., with intonational stress) something in her communicative act as out of the ordinary, so that the recipient will not make the normal inference but rather a different one. For example, Liebal et al. (2011) had an adult and a two-year-old child again tidying up toys into a large basket. In the normal course of events, when the adult pointed to a medium-sized box on the floor, the child took this to suggest that she should tidy up this box into the basket as well. But in some cases the adult pointed to the box with flashing eyes and a kind of insistent pointing directed at the child, obviously not the normal way of doing it. The adult clearly intended something different from the norm. In this case, many children looked at the adult puzzled but then proceeded to open the box and look at what was inside (and tidy *it* up). The most straightforward interpretation of this behavior is that the child understood that the adult was anticipating how she would construe a normal point, which he did not want, and so he was marking his pointing gesture so that she would be motivated to search for a different interpretation. This is the child thinking about the adult thinking about her thinking about his thinking.

And so, the kind of thinking that goes on in human cooperative communication is evolutionarily new in that it is perspectival and socially recursive. Individuals must think (simulate, imagine, make inferences) about their communicative partner thinking (simulating, imagining, making inferences) about their thinking—at the very least. Great apes show no signs of making such inferences, and their failure to comprehend even the simplest acts of cooperative pointing, for example, in the object choice task (while making non-recursive inferences in the same task setting), provides positive evidence that they do not. Human thinking in cooperative communication also involves a new kind of social self-monitoring, in which the communicator imagines what perspective the recipient is taking, or will take, on his intentions toward her intentions—and so imagines how she will comprehend it. In all, what we have at this point in our evolutionary story of human communication is individuals attempting to coordinate their intentional states, and so their actions, by pointing out new and relevant situations to one another. This relies on their having a certain amount and type of common ground, and it requires, further, that the interactants make a series of interlocking and socially recursive inferences about one another's perspectives and intentional states.

### Symbolizing in Pantomime

Beyond the pointing gesture, the second form of "natural" communication that humans employ is spontaneously generated, nonconventional iconic gestures, or pantomime. These gestures are used to direct the imagination of others to nonpresent entities, actions, or situations. Iconic gestures go beyond simply directing attention to situations deictically, as in pointing, by actually symbolizing an entity, action, or situation in an external icon. Iconic gestures are "natural" because they employ normally effective intentional actions, just in a special way. The recipient can, on the basis of observing them, imagine the real actions or objects the communicator is pantomiming, and then, in the context of their common ground, make the appropriate inference to his communicative intention. Examples of informative uses of iconic gestures would be things like warning of a nearby snake by moving one's hand in a slithering motion, telling of a deer at the waterhole by miming antlers on one's own head (or the sound of his vocalization), or identifying the whereabouts of a friend by pantomiming him swimming. With the appropriate common ground, such gestures communicate very effectively about all kinds of nonpresent situations.

No nonhuman primates use iconic gestures or vocalizations. Great apes could easily gesture with their hands the way that humans do to mime eating or drinking, but they do not.[8] Indeed, great apes do not even understand iconic signs. In a modified object choice experiment, a human held up a replica of the object under which food was hidden. Two-year-old children knew that this meant to search in the similar-appearing object, while chimpanzees and orangutans did not (Tomasello et al., 1997; Herrmann et al., 2006). In ongoing research we have been trying to elicit iconic gesturing from apes in situations in which it would benefit them to do so (e.g., showing a human how to extract food for them from an apparatus that only they know how to operate), but so far with no success. Presumably, great apes do not understand iconic gestures because they do not understand communication marked ostensively as "for you" (cooperatively). If an ape views someone hammering a nut, they know perfectly well what he is doing, but if they view him making a hammering motion in the absence of any stone or any nuts, they are simply perplexed. To comprehend iconic gestures, one must be able to see intentional actions performed outside of their normal instrumental contexts as communication—because they are marked as such by the communicator via various kinds of ostensive signals (e.g., eye contact). Extending an analogy from Leslie (1987) on pretense, the bizarre action must be "quarantined" from straightforward interpretation as an instrumental action by marking it as "for communication only."

Another prerequisite for an individual to produce an iconic gesture is that he is able to produce with his body an action that "resembles" a real action (or object). Presumably the ability to do this derives from the ability to imitate, at which humans are especially skillful compared with other apes (Tennie et al., 2009). Somehow early humans came to understand that "imitating" an action not for real but with an ostensive communicative intention (simulating it in action) could lead a recipient to imagine all kinds of referential situations not in the current perceptual scene. One potentially important social context in this connection is teaching, which has the evolutionary advantage that the primal scene is one of an adult instructing its offspring. Csibra and Gergely (2009) explicate what they call "natural pedagogy" and note its close connection to cooperative communication. The most basic form of natural pedagogy is demonstrating: showing someone how to do something by either doing it directly or pantomiming it in some way. And like communication, the action is being done not for its own sake but for the benefit of the observer/learner. Communicating with iconic gestures thus requires both an understanding of ostensive communication and some ability to imitate actions.

Importantly, iconic gestures may depict the referential object or action quite faithfully, but then it can still be, as in pointing, a large inferential leap to the underlying communicative intention. Thus, to bridge the gap, just as in the case of pointing, common ground and mutual assumptions of cooperation and relevance are needed. If I mime for you a snake's motion as we approach a cave, if you do not know that snakes are often found in caves, you might wonder why I am waving my hand in that way. In the contemporary world, we recently observed a young child going through airport security. The guard scanning her with his wand moved his hand in a circular motion to tell her to turn around so he can scan her backside. Staring at him, she slowly began moving her hand in a circular motion back at him—she did not understand that his hand was meant to represent her body. Apparently, they did not have in their common ground airport security procedures.

Whereas there is only one basic pointing gesture,[9] there are myriad possible iconic gestures—a "discrete infinity," perhaps. With iconic gestures there is, or at least can be, a more or less one-to-one correspondence of gestures and the intended referent (though typically only one aspect of an intended referential situation is mimed). This means that iconic gestures, even though not conventional, have a kind of semantic content. With pointing, I can in principle indicate for you the shape, size, or material of a piece of paper, within the appropriate common ground, but the unique perspective in each case is not in any way contained "in" the protruding finger itself (see Wittgenstein's [1955] incisive, if cryptic, discussion of this issue). But with iconic gestures, I would indicate for you the shape, size, or material of a piece of paper—or whether I want you to write on the paper or throw it away—by depicting each of these different aspects or actions with different icons. The momentous new feature of iconic gestures is thus that the different perspectives of things and situations only implicit in pointing are now expressed overtly in external symbolic vehicles with semantic content.

Relatedly, the vast majority of communicative conventions in a natural language are category terms. That is, common nouns and most verbs are conventionalized for reference to categories of entities such as *dog* and *bite*, which means that to make reference to a specific dog or instance of biting, we must do some kind of pragmatic grounding (such as with *the* or *my dog*, or *the dog who lives next door* in the case of nouns; or tense and aspect markers, as in *is biting* or *bit*, in the case of verbs). Iconic gestures are already category terms, because they implore the recipient to imagine something "like this." (It is possible that one could iconically gesture an individual as well—for example, by mimicking her idiosyncratic mannerisms—and so the distinction between common and proper nouns is at least in principle possible in this modality.) The categorical dimension is bound up with perspective in the sense that calling someone either Bill or Mr. Smith is not perspectival because these are not category terms, but calling him a father or a man or a policeman is perspectival because it puts him "under a description," that is, it "perspectivizes" him differently on different occasions for different communicative purposes.

Iconic gestures are thus an important step on the road to linguistic conventions in that they are symbolic, with semantic content, and are at least potentially categorical. An interesting fact that reinforces this point is that although young children produce some iconic gestures from early in development, they actually go down in frequency over the second year of life as children begin learning language, whereas pointing increases in frequency during the same period. One hypothesis is that pointing increases because it does not compete with language but complements it by performing a different function. As symbolic vehicles with semantic content, iconic gestures compete with linguistic conventions, and they lose the competition—for many obvious reasons—which usurps the need to create spontaneous gestures on the spot, except in a few exceptional circumstances. If one imagines an evolutionary

## BOX 2. Pantomime as Imagining in Space

Communicating with iconic gestures and pantomime might plausibly have had two large and important cognitive consequences. The first stems from its intimate involvement with—and therefore stimulation of—imagination and pretense. Iconic gestures enable reference to things ever farther removed in space and time than does pointing, and at the moment of communication these must be imagined. When I gesture to inform you that there is an antelope out of sight over the hill, or to warn you that there are snakes in this cave we are about to enter, or to relate what happened to us on our just-finished hunting trip, I have to act out whole scenarios in which some of the key players are not even present and the actions are either long past or only predicted.

The proposal, then, is that iconic gesturing both depends on preexisting skills of imagination and also takes these skills to new places. Whereas a chimpanzee might imagine what awaits her at the waterhole, we are now talking about depicting for another person in some kind of playacting such imagined scenes—tailored for the recipient's knowledge and interests, given our common ground, so that she will be capable and motivated to comprehend. It is not unreasonable to suppose, then, that humans evolved ever more powerful forms of imagination to be able to act out scenes for others—in a kind of joint imagining. Indeed, we see this behavior in very young children on a daily basis as they pretend with a parent or peer that this stick is a horse or that they are Superman. The evolutionary origins of pretend play—which would seem to be a bit mysterious because its function is not so obvious—are thus, in the current account, to be found in pantomiming as a serious communicative activity. In modern humans, pantomiming for communication has been supplanted by conventional language. As children learn a conventional language, their tendency to communicate to others by creating pretend scenarios in gesture has no place to go, so to speak. They thus play with this ability and create pretend scenarios together with others, as a pretense activity with no other motivation. A number of scholars have argued as well that engaging in pretense is one source for the distinction between appearance and reality (e.g., Perner, 1991), as I act out X fictively in order to represent the real X, as well as for counterfactual thinking in general (Harris, 1991).

*(continued)*

BOX 2 (*continued*)

And so the first surprising effect of iconic gestures is that their emergence in human evolution led to skills of acting out pretend scenarios with and for others, which may be the basis for humans' creation of all of the "imaginary" situations and institutions within which they reside. In addition, to anticipate our story a bit, it is also reasonable to suppose that the creation of what Searle (1995) calls cultural "status functions" such as being a president or a husband—and pieces of paper standing for (indeed, constituting) money—has its phylogenetic and ontogenetic roots in pretend play in which children together anoint a stick as a horse, which gives the stick special powers, in a manner very similar to anointing a person as a president (Rakoczy and Tomasello, 2007). If thinking is at base a form of imagining, then one can hardly overestimate the importance of imagining things for other people, as embodied in iconic gestures, for the evolution and development of uniquely human thinking (Donald, 1991).

The second cognitive effect of iconic gestures and pantomiming is even more speculative. Almost everyone who studies human cognition recognizes the crucially important role of spatial conceptualizations. There are undoubtedly multiple reasons for this, some of them having simply to do with the importance of space in primate cognition in general. It is well known, for example, that episodic memory has intimate connections to spatial cognition.

But more recently, some theorists have dug more deeply into this connection. Beginning with the pioneering work of Lakoff and Johnson (1979), it is well known that humans quite often talk about abstract situations or entities metaphorically or analogically in terms of concrete spatial relationships. As just a few examples, we talk about putting things into and taking them out of our lectures, we fall into love, we are on our way to success, or we are going nowhere in our career, or I am out of my mind, or she is coming to her senses, and on and on. We are not talking about just surface metaphors here, but very basic ways of conceptualizing complicated and abstract situations. Thus, in his follow-up work, Johnson (1987) identified a number of so-called image schemas that seem to permeate our thinking, such as *containment* (in and out of a lecture), *part-whole* (the foundation of our relationship), *link* (we are connected), *obstacle* (my lack of education gets in the way of my social life), and *path* (we are on our way to marriage).

Even in the grammars of languages, a number of scholars have noted the inordinate prominence of space, with some even creating such things as "space grammars." Some of the early work on syntactic case relationships also emphasized that many case markers emerged first historically from words for spatial relationships (adpositions of various kinds). Talmy (2003) has posited a human imaging system that structures grammar via a very strong spatial component. Thus, one of his central schemas is the force dynamic schema in which actors cause effects in other entities (e.g., investors' anxieties crashed the stock market), and another is various kinds of fictive motion along paths. He has also noted that many complex relationships are expressed spatially, with topological relationships predominating. Even more strongly, conventional signed languages use space to depict all kinds of grammatical relationships, from anaphoric reference to case role (e.g., Liddel, 2003)—which is important if humans' earliest linguistic conventions were, as hypothesized here, in the gestural modality.

In terms of ontogeny, Mandler (2012) has argued and presented evidence that children's earliest language is made possible by a set of mainly spatial image schemas such as animate motion, caused motion, the path of motion, obstacles to motion, containment, and so forth. These form the conceptual foundation for children's early talk about agents doing things (Slobin's [1985] manipulative activity scene) and objects going places (Slobin's [1985] figure-ground scene, in which objects move along paths). These are the things children first talk about, and fundamental spatial relationships in motion along paths play a prominent role at all stages.

The speculation is thus that in addition to numerous other reasons for space being important in human cognition, a critically important reason is that at an early stage in their evolution humans conceptualized many things for others in their gestural communication in a fictive space with fictive actors and actions. Basically, the only way to depict many things in spontaneous, nonconventionalized gestures is by acting out in space the referent objects and events. And so if we believe that human thinking is intimately tied to communication—how we have come to conceptualize things for others—then the fact that we did this for some time in our history by pantomiming in space may go a long way toward explaining the inordinately important role of space in human cognition.

analog, the story would be one of conventional forms of communication mainly taking over from iconic gestures, whereas pointing would persist. In both evolution and ontogeny, then, the ability to act out nonactual situations could then potentially emerge again in other functions, for example, pretense and other forms of fiction (see box 2).

Iconic gestures thus represent a kind of middle stage in human communication and thinking that bridges from pointing for others informatively and perspectivally in the context of common conceptual ground to conventional linguistic communication. This bridging step involves external forms of symbolic representation that have semantic content categorized. Nevertheless, iconic gestures almost always have a potentially problematic ambiguity of perspective. If I mime the throwing of a spear, who is supposed to be throwing it—me, you, or someone else? Of course, this typically is determined through our common ground context; it is normally clear if I am requesting you to do it, expressing my desire to do it, or reporting on our friend's activity. But in some situations—for example, depicting the morning's hunting events—it might not be clear who is throwing the spear. The only way to resolve this ambiguity is with further communicative acts, either deictic or iconic. And this leads us to the most complex way in which early humans communicated with natural gestures before conventional languages: combining their gestures in multiunit expressions.

### Combining Gestures

Great apes do not create new communicative functions by combining their gestures, their vocalizations, or their gestures and vocalizations together (Liebal et al., 2004; Tomasello, 2008). But humans do, including young children from the earliest stages of their communicative development, and including even children exposed to no conventional language, vocal or signed, at all (Goldin-Meadow, 2003).

While there is no principled reason why someone could not string together various pointing gestures—and individuals may do this on occasion—this is not commonly observed. Beginning language learners combine their earliest linguistic conventions with pointing or other conventions, and beginning sign language learners produce iconic or conventional signs in combination with pointing (as do, again, children exposed to no conventional language at all; Goldin-Meadow, 2003). As originating context in evolution, one can easily imagine situations in which an individual pantomimed something—such as eating—and then immediately afterward, in a second thought, pointed to some particular piece of food—for instance, the fruit over there (this process is thus analogous to the "successive single word utterances" in early child language, or the "broken" utterances in pidgin languages). But then, through a process of "mental combination" (Piaget, 1952), these successive thoughts or intentions came to be integrated into a single thought or intention and so expressed as a single utterance within a single intonation contour. With some minimal skills of categorization, individuals could form a schema comprising, for example, an iconic gesture for eating followed by indexical indication of anything edible either by oneself or by others. Productivity in thinking would be thus scaffolded and enhanced by this overt communicative schema.

It is important to emphasize that just as in child language, in early human communication there would have been functional continuity between different expressions of the same referential intention, no matter their internal complexity. For example, one could communicate that there are snakes in the cave either with a snake motion as we approach the cave or by combining a snake motion with a pointing gesture to the cave (e.g., if we are not approaching it)—both with the same communicative intention or function. Combining symbolic and deictic vehicles is not the creation of new communicative intentions, primarily, but rather the parsing of existing ones into their component parts. This means that in combinations a single gesture is typically indicating only one aspect of a situation. Thus, whereas the snake motion while approaching the cave is intended to communicate that there are snakes in the cave, in combination with pointing to the cave (or iconically depicting the cave) this snake motion now only indicates the snakes themselves, as the rest of the situation is indicated through other communicative devices. This focus on function and the parsing of situations into components with different subfunctions are responsible for the hierarchical organization of human communication.

With gesture combinations we now also have the possibility of beginning down the path to the subject-predicate organization characteristic of full propositions.[10] Two ingredients are involved, both of which are already present, in nascent form, in pointing inside collaborative activities. The first ingredient is the specific cognitive distinction between events and participants. Even apes learning human-like forms of communication distinguish events and participants in their sign combinations (see Tomasello, 2008, for evidence). The second ingredient is the distinction between shared (or given) and new information. As noted above, even in pointing there is an implicit distinction between the shared common ground, which typically is not indicated specifically by the pointing gesture, and the new and noteworthy situation, which is indicated deictically. But this is all implicit. With gesture combinations, what often happens is that one or more signs are used to make contact with the common ground—typically to use it as a perspective or "topic"—and then to indicate with another sign the new and interesting information. In many situations, one can imagine that one points to a perceptually present referent—to make sure that it is shared—and then iconically signs about some aspect of it that one thinks is new and noteworthy for the recipient.

The overall picture is thus that early humans used their pointing and iconic gestures, both singly and in combination, to communicate much more richly and powerfully than did their primate cousins. This new form of communication took place initially inside of collaborative activities, which supplied the interactants with both the necessary common conceptual ground and the necessary opportunities for interchanging roles and perspectives with their partner. Early humans' cooperative communication with natural gestures thus required both levels in our dual-level conception of joint collaborative activities: joint goals and attention, as the shared aspect, and individual roles and perspectives, as the individual aspect. And none of this required language. Communicators conceptualizing or perspectivizing things in different ways for different communicative partners (depending on judgments of common ground, relevance, and newness), and then recipients comprehending the intended perspectives through socially recursive inferences, is not the result of becoming a language user, but rather its prerequisite.

## Second-Personal Thinking

We are trying to get to the full flowering of modern human objective-reflective-normative thinking in the context of culture and language. We are halfway there. With the reconstructed early humans we are picturing here, we have creatures who are not just strategizing how to obtain food or mates in bigger, better, and faster ways than others, as are great apes, but rather who are attempting to coordinate their actions and intentional states with others via evolutionarily new forms of collaborative activity and cooperative communication. They are not just organizing their actions via individual intentionality; they are also organizing them via joint intentionality. And this changed the way they imagined the world so as to manipulate it in acts of thinking.

### Perspectival, Symbolic Representations

Great apes schematize cognitive models for the various types of situations that are recurrent and important in their lives. And so when early humans began engaging in obligate collaborative foraging, they schematized a cognitive model of the dual-level collaborative structure comprising a joint goal with individual roles and joint attention with individual perspectives. With cooperative communication, early human individuals began overtly indicating or symbolizing for a partner situations that were relevant to her, given her individual role and perspective in the joint activity. To do this, they created evolutionarily new forms of natural gestures—pointing and pantomiming—whose use resulted in cognitive representations that had three new and transformative characteristics.

PERSPECTIVAL. Conceptualizing things from different perspectives comes so naturally to humans that we consider it as almost inevitable; it is just the way cognition works. Typically in cognitive science, concepts are characterized in English words, such as *car*, *vehicle*, and *anniversary present*, that can be applied as needed, even to the same entity sitting in the driveway. But this way of doing things is not inevitable; indeed, it is not even possible for creatures that cannot "triangulate" with another individual simultaneously on the same entity. Great apes may sometimes apply different schematic representations to one and the same entity: on one occasion a particular tree is an escape route, whereas on another occasion it is a sleeping place. But each of these different conceptualizations is tied to the individual's current goal state; she may know many things about the tree, but she does not entertain them as alternative possible construals simultaneously, and so they are not interrelated perspectives in the way we have defined them here (and this is true even if the ape is solving a problem by imagining nonactual entities or situations, because even here she is doing this aimed only at her current problem situation).

In contrast, when early humans began to communicate cooperatively with others, they were constantly taking the other's perspective on a situation or entity to which they themselves were already attending (they were triangulating with the other). Indeed, each time they communicated, they had to make their communicative act relevant and new for the recipient in the context of her goals and values, their common ground, and her existing knowledge and expectations. As they were thus thinking how their communicative act might fit into the life of the recipient, communicators had to consider several alternative perspectives simultaneously, and only then choose a communicative act to instantiate one of them. For example, in order to warn of danger, they might pantomime for a recipient as they approached a cave either a snake, or a snake bite on the leg, or just a general danger sign (which the recipient would know, in the context of their common ground about caves, meant snakes).

The key point from the perspective of cognitive representation is that communicators were not tied to their own goals and perspectives, but rather they were considering alternative perspectives for another person, whose conative and epistemic states they could only imagine. For her part, the recipient, in order to make the abductive leap necessary to grasp the communicator's communicative intentions, had to then simulate his perspective on her perspective (at least). This transacting in perspectives meant that early human individuals did not just experience the world directly for themselves, in the manner of all apes but, in addition, at least in some aspects, experienced the exact same world viewed simultaneously from different social perspectives. This triangulating process inserted for the first time a small but powerful wedge between what we might now call the subjective and the objective.

SYMBOLIC. Iconic gestures, or pantomime, also seem mundane to humans, who have the ability to imitate one another's actions, and to even imitate or simulate their own past actions outside of their normal instrumental context. But they are anything but mundane, as they represent the first acts by any primate (arguably any animal species) that attempt to *re*-present for a recipient, in overt action, some event or entity, so that she will imagine it. Iconic gestures also require that the recipient comprehend communicative intentions (in our story, already in the comprehension of pointing) so that she can "quarantine" these gestures as not actual instrumental acts but rather acts of communication.

Producing communicative acts that resemble their intended referents (e.g., miming a monkey climbing) creates a symbolic relationship in which the act is meant to evoke in imagination the intended referent (e.g., a monkey or an act of climbing or a monkey climbing), which is hoped to lead the recipient to infer the communicator's communicative intention (e.g., that they go hunting for monkeys now). Like pointing, iconic gestures perspectivize a situation, but unlike pointing, they do this articulately in the symbolic vehicle itself. For example, with iconic gestures one would have different icons for "monkey" and "food" even if, on different occasions, they were used for the exact same animal, whereas in pointing, the act (protruding finger) would be the same in both cases, with the common ground of the collaborative activity (whether we are admiring the local fauna or seeking sustenance) carrying the semantic weight. Another important feature of iconic gestures is that they are mostly categorical in nature, that is, used to conceptualize or perspectivize things, events, or situations "like this." In choosing what to act out for others in pantomime, then, communicators construe the situation from a particular perspective categorically, as opposed to other possible categorical perspectives.

QUASI-PROPOSITIONAL. Combining gestures into a single communicative act parses the referent situation into something like event participant structure, which limits the semantic scope of each gesture. Thus, pantomiming a monkey in combination with pointing to a spear now suggests even more articulately the desired hunting trip, but the gesture for monkey is now confined to symbolizing the monkey only, not the hunting trip as a whole. In combination with the already established tendency to background knowledge in common ground (topic) and to explicitly communicate about new information (focus), this parsing creates a nascent subject-predicate organization in the communicative act—resulting in something on the way to full propositions. (Interestingly, great apes raised by humans with a human-like communication system typically make the event-participant distinction, but not the topic-focus distinction [because they do not have any notion of a shared focus of attention or topic], and so they do not have subject-predicate organization in their multiunit communicative acts [see Tomasello, 2008].) The addition of a new cooperative communicative motive made for two distinctly marked motives (requestive and informative), which created a first nascent distinction between communicative force and content.

With the advent of early human collaboration and cooperative communication, then, the cognitive representation of experience in type-token format, as in great apes, was "cooperativized." Individuals interacting in joint attention and common conceptual ground could conceptualize the same event, entity, or situation simultaneously from multiple perspectives. The symbolization of these perspectives in categorical iconic gestures and in gesture combinations with topic-focus organization, with some indication of a force-content distinction, made them at least incipiently propositional as well. This process might be seen to effectively decontextualize (cooperativize or make less egocentric) the individual's experience of the world, as he decides under which symbolic description to represent a situation for a communicative partner. With this perspectival crack in the experiential egg, we are on our way to thinking that is in some sense "objective."

### Socially Recursive Inferences

Socially recursive inferences, once again, seem so natural for humans as to be barely noticeable: I wonder what she thinks I'm thinking. Great apes make inferences about experience—they simulate the causes and outcomes of physical and social situations—but they do not make inferences about what the other is thinking about their thinking. Such inferences begin with early humans attempting to coordinate their actions and attention with others in collaborative activities with joint goals and attention, but they come into full flower with early humans attempting to coordinate their intentional states and perspectives with others in cooperative communication.

In the context of a joint collaborative activity, early human communicators began thinking about (i.e., simulating) how best to formulate their communicative act for a recipient, with the goal of both honesty (engendered by a concern with being cooperative in general) and communicative efficacy. The concern with honesty—especially given that recipients were now becoming "epistemically vigilant" (Sperber et al., 2010)—puts us on the road to a commitment to the truth of our communicative acts. The concern with communicative efficacy required that both communicator and recipient anticipate the perspective of their partner, which required socially recursive inferences that embedded the intentional states of one partner within those of the other. In addition, the production of overt combinations of gestures for others, once they were schematized, created unprecedented new possibilities for productive inferences about nonactual or even counterfactual states of affairs. Early human inferences thus display two new and transformative properties.

SOCIALLY RECURSIVE. We may reasonably ask why early human communicators began making socially recursive inferences in the first place. The short answer is that they assumed together in common ground that the communicator had cooperative motives, and so they were collaborating toward the joint goal of recipient comprehension. In this context, they were each trying to help the other—as in all joint collaborative activities—and this meant simulating what the other was thinking about their thinking. And since pointing and pantomiming on their own are quite weak communicative vehicles, inferential leaps of at least some distance are always required to reconstruct the communicator's communicative intention—so that at least some help is almost always needed.

And so developed a form of communication in which a communicator intended that a recipient know something, for her benefit. The recipient understood this and so, for example, understood that he *intends* for me to *know* that the banana is in that bucket. The communicator, for his part, knew that the recipient would make such an inference if he helped her to do so by alerting her to the fact that he had such an intention (the Gricean communicative intention that the recipient notice that the communicator wants her to know something). This may not be one multiply embedded communicative intention, as in the Gricean analysis, but rather, as argued by Moore (in press) two singly embedded intentions: I intend that you notice that this communicative act is for you, plus I intend that you know that the banana is in that bucket. Nevertheless, the single embedding in this second intention is already more than great apes can do, and so it represents a new form of recursive inference (the production version occurring when the communicator simulated the recipient's intentional states in order to formulate communicative acts that would be readily comprehensible for her—not throwing the ball *at* her, but rather *to* her).

COMBINATORIAL. Communicating with others using overt gestures, and especially being able to combine gestures to communicate with others in more complex ways, enabled new processes of productive thinking. In their natural communication with one another, great apes do not combine gestures with one another (nor vocalizations) to communicate something new. Their thinking is thus confined to imagining novel situations using their past individual experiences reconfigured in new ways. But once early humans began imagining situations from the perspective of the other in order to communicate with combinations of gestures, and then schematized these combinations, they had the possibility to go beyond their own experience to think about something that others might experience, or even something impossible. For example, I might produce an iconic gesture for traveling followed by pointing to a location, which I generalize to any location. I might then imagine or communicate, via this schema, our child traveling to the sun—something I consider causally impossible. When humans began schematizing communicative constructions with abstract slots in this way, they created for themselves almost unlimited combinatorial freedom. Schema formation in communicative acts and the parsing of communicative intentions into discrete overt components represent a significant step in the direction of the kind of "inferential promiscuity" characteristic of modern human thinking in a conventional language.

Beyond the new possibilities for creating novel, even counterfactual thoughts via external communicative vehicles, a number of theorists have emphasized the necessary role of such external vehicles for individuals to reflect on their own thinking (e.g., Bermudez, 2003). When individuals formulate an overt communicative act and then perceive and comprehend it as they produce it, they are, in effect, reflecting on their own thinking (a process that may become internalized so that we may think about things that we could potentially communicate overtly). Because the gesture combinations at this point have only limited semantic content (e.g., no logical vocabulary and no propositional attitude vocabulary), early humans could reflect only in a highly limited way on their own thinking.

With the advent of early human collaboration and cooperative communication, then, the causal inferences of great apes were, like their cognitive representations, "cooperativized." This meant that the communicator's inferences were about what was the situation from the perspective of the recipient, and the recipient's inferences were about the communicator's simulations of her simulating his perspective. Overt combinations of symbols, especially if schematized, led to the possibility of thinking various new and even counterfactual thoughts, as well as to the first, rather modest, reflections on one's own thinking. With all of these new inferential possibilities, then, we are now well on our way to thinking processes that are truly reflectively reasoned.

## Second-Personal Self-Monitoring

Great apes self-monitor their goal-directed behavior, including its psychological underpinnings with respect to such things as memory and decision making. But great apes are not normative creatures. They experience "instrumental pressure," for example, when they have a goal to eat food and they know that food is available at location X; this implies that they "must" go to location X. But this is just the way control systems with individual intentionality work: a mismatch between goal and perceived reality motivates action. In contrast, early humans began to self-monitor from the perspective of others and, indeed, self-regulated their behavioral decisions with others' evaluations in mind. Now we may talk of something that is socially regulated, that is, socially normative, albeit only in second-personal (as opposed to agent-neutral) form. There were two manifestations.

COOPERATIVE SELF-MONITORING. First, because the collaborative activities of early humans were interdependent and operated with partner choice, each individual, even the most dominant, had to respect the power of other individuals, even the most subordinate, to exclude them from collaborative opportunities. Early humans thus developed not only an ability to make evaluative judgments about others' cooperative proclivities but also an ability to simulate, and so to anticipate, the evaluative judgments that others were making about them. Young human children are concerned with the social evaluations of others from the preschool years on as they attempt to actively manage the impression they are making on them (Haun and Tomasello, 2011), but chimpanzees seem to be not so concerned (Engelmann et al., 2012).

Early humans' concerns for how their collaborative partners viewed them—and their active attempts to manage this impression—provided a new motive for actions, namely, to coordinate with the evaluative expectations of potential partners. Individuals thus began to cede power over themselves to the second-personal evaluations of others because these evaluations determined their future collaborative opportunities. From the point of view of normativity, this meant that in making their behavioral decisions, humans not only experienced individual instrumental pressure but also experienced second-personal social pressure from their partners in social engagements. This constitutes, in the current account, one origin of what will later become social norms of morality.

COMMUNICATIVE SELF-MONITORING. Second, because early human communicators wanted to facilitate the recipient's comprehension, they had to actively self-monitor their potential communicative acts in anticipation of how they might be comprehended and/or interpreted by the recipient. They thus engaged in a self-monitoring of the communicative process, from the perspective of the recipient, especially for intelligibility.

Mead (1934) pointed out the key role of overtness here. As they communicated with others in overt acts—either deictic or symbolic—early humans saw or heard themselves performing those acts, in which case they then comprehended them (as perspectivized for another) as the recipient. Communicators thus adjusted their communicative act so as to maximize the comprehension of the recipient, as part of their commitment to the collaborative act of cooperative communication. Making such adjustments required self-monitoring and evaluating communicative acts for comprehensibility from the perspective of specific communicative partners, each with her own individual knowledge and motives and common ground with the communicator. This constitutes, in the current account, one origin of what will later become social norms of rationality.

The early humans we are picturing here would thus have been able to engage in two kinds of cooperative and communicative self-monitoring that great apes cannot—because great apes do not engage in the kind of joint collaborative activities and cooperative communication that engenders such social self-monitoring. Early humans simulated the evaluative judgments that others made of them with regard to their cooperative proclivities—precursors to norms of morality—and also with regard to the intelligibility of their communicative acts—precursors to norms of rationality. Importantly, the evaluations we are talking about here come from particular individuals, and so we are still a way from the kind of agent-neutral, "objective" norms by which modern humans evaluate others and themselves. But we have begun the process of socially normativizing individual thinking.

## Perspectivity: The View from Here and There

It is now widely accepted that what most clearly distinguishes nonhuman primates from other mammalian species cognitively is their complex skills of social cognition. Dunbar (1998), for example, documented that primate brain size correlates most strongly not with primates' physical ecology but, rather, with their social group size (as a proxy for social complexity). But primates' special skills of social cognition are aimed mainly at competition (i.e., their intelligence is Machiavellian), in which they keep track of all of the various dominance relationships in the group, as well as all of the various affiliative relationships in the group as these might affect the competition for food and mates.

The question thus arises whether the uniquely human skills of cognition and thinking that we have identified here might instead have arisen for competition. On the level of evolutionary function (ultimate causation), this is true almost by definition, since evolutionary success is defined as having more offspring than others. But on the level of proximate mechanism, we do not think it likely that perspectival cognitive representations, socially recursive inferences, and social self-monitoring could have arisen directly out of competitive contexts. It is true that, in theory, one could get a kind of arms race in mind reading to deal with competitive situations. In competition, individuals could come to realize that both me and my competitor are focused on the same resource at the same time (joint attention?) and then to try to outcompete her for that resource by thinking about what she is thinking about my thinking. But what we for sure cannot get strictly from competition are the unique forms of cooperative communication in which humans engage. Unlike other primates, humans use their communicative acts to actually encourage others to discern their thinking. Thus, human communicators take the perspective of others in order to determine their goals and interests so that they can then inform them of something helpful to them. Those others want this helpful information, and so they do their best to help the communicator discern their goals and interests, and also to discern their knowledge and expectations so that he may formulate his communicative act in a comprehensible manner. Humans, but not other primates, thus collaborate in their communication to make it easier for the other to take their perspective and even to manipulate it if they so desire.

An especially enlightening example of a similar cooperative process concerns a unique physical characteristic of humans. Of the more than two hundred species of primates, only humans have highly visible eye direction (because of especially visible white sclera; Kobayashi and Koshima, 2001). And only humans use this information. Thus, when tested in various conditions contrasting head and eye direction, twelve-month-old human infants tended to follow the eye direction over the head direction of others, whereas great apes tended to follow the head direction of others only (Tomasello et al., 2007b). For humans to have evolved conspicuous cues of gaze direction, there must have been some advantage to the individual to "advertise" her eye direction for others. This suggests predominantly cooperative situations in which the individual may rely on others using this information collaboratively or helpfully, not competitively or exploitively. The point is that human communicative acts serve to advertise the internal states of individuals in this same way, and so it also suggests cooperation of this same type (e.g., cooperative requests such as "I'd like some fruit" are "advertisements" of my internal state of desire, and informative utterances such as "There is some fruit over there" are public offers of helpful information). Communication of this type could never be adaptively stable in contexts that were not fundamentally cooperative, and so fully human-like skills of joint intentionality could never evolve solely in the context of competition.

There can be no doubt that the last common ancestor to humans and other primates engaged in individual thinking in pursuit of individual goals, mostly in order to compete with groupmates for valued resources. Along the way, they attended to situations relevant to those goals. Early human individuals—in response to a changing feeding ecology—then began to join together with other individuals dyadically in pursuit of joint goals, and they jointly attended to situations relevant to that joint goal. Each participant in the collaboration had her own individual role and her own individual perspective on the situation as part of the interactive unit. This dual-level structure—simultaneous jointness and individuality—is the defining structure of what we are calling *joint intentionality*, and it is foundational for all subsequent manifestations of human shared intentionality.

The problem was how to coordinate these collaborative activities as they became ever more complex, both to negotiate a joint goal and to coordinate the two different roles. The solution was cooperative communication. Early humans directed the attention of their collaborative partner to relevant situations by pointing, which required taking her perspective and simulating her thinking (i.e., in terms of the abductive leap she might be expected to make given different possible communicative acts). To comprehend, the recipient had to take the perspective of the communicator taking her perspective— which constituted to a new form of socially recursive inferring. Early humans' concern that their partner comprehend them led to social self-monitoring via the anticipated evaluations of the partner with respect to the comprehensibility of the communicative act.

The basic cognitive challenge in all of this was to coordinate one's own perspective with the perspective of one's collaborative partner. And so, as early humans engaged in the truck and barter of making a living collaboratively, they began to truck and barter in perspectives with their interactive partners communicatively—and in their own perspectives reflectively to some degree— and this gave human cognitive representation and inference a new kind of flexibility and power. Now, instead of just their own view on the world, early humans could also view the world at the same time from the perspective of the other, which might also include her perspective on my perspective. Early humans had not just a great ape view from here, but rather a view simultaneously from here and there.

We do not know precisely who these early humans were, but we may speculate *Homo heidelbergensis* some 400,000 years ago, living as loosely structured bands or pools of recurrently collaborating partners. Of course *Homo heidelbergensis* did not engage in modern human forms of fully objective-reflective-normative thinking. Their thinking was not "objective" but rather was still tied to the two second-personal perspectives of "I" and "you." Their thinking was only weakly reflective because they could express very few of their intentional states or cognitive operations externally in communicative vehicles (and so they could act as both producers and comprehenders of only some limited semantic content). And their thinking was socially normative only in the sense that they were concerned with how their partner evaluated their cooperative behavior and comprehended their communicative acts, not with the group's normative standards. There is thus no question that we are still some way from modern human collective intentionality and its objective-reflective-normative thinking. But, we would argue, the "in-between" step of early human joint intentionality and its perspectival-recursive-socially monitored thinking was necessary for getting there. It was necessary because the transition to modern humans was all about creating cultural conventions, and if these were to be in a cooperative direction—as they almost invariably were— then some very strong cooperative tendencies had to be already present in the individuals doing the conventionalizing.

Together, then, early human collaborative activities and cooperative communication represent a kind of second-personal "cooperativization" of great ape lifeways and thinking. But these evolutionarily new forms of second-person social interaction involved joint engagement with specific other persons on specific occasions only, and they did not retain their special characteristics very far outside of the collaborative activities themselves. And so, despite the great leap forward represented by this new joint intentional way of living, communicating, and thinking, the next leap forward will have to take this "cooperativized" cognition and thinking and "collectivize" it by conventionalizing and institutionalizing—and so normativizing and objectifying—almost everything.

# 4

## Collective Intentionality

There is no thinking apart from common standards
of correctness and relevance, which relate what I *do* think
to what *anyone* ought to think. The contrast between
"I" and "anyone" is essential to rational thought.

—WILFRID SELLARS, *PHILOSOPHY AND THE SCIENTIFIC IMAGE OF MAN*

A modern human society may be characterized in two dimensions. The first is its *synchronic* social organization: the coordinated social interactions that make it a society in the first place. Early human individuals, as we have seen, coordinated in acts of collaborative foraging with specific others from a loosely structured pool of collaborators. But now with modern humans we need to scale up to much larger social groups with much more complex social organization, that is to say, to fully cultural organization. Modern humans became cultural beings by identifying with their specific cultural group and creating with groupmates various kinds of cultural conventions, norms, and institutions built not on personal but on cultural common ground. They thus became thoroughly group-minded individuals.

The second dimension of a modern human society is its *diachronic* transmission of skills and knowledge across generations. Social transmission of one sort or another was almost certainly important in the lives of early humans (as it is in some apes' lives), as difficult-to-invent, tool-based, subsistence activities became ever more complex and important for survival. But now with modern humans we need to scale up to full-fledged cultural transmission supporting cumulative cultural evolution. This required that modern humans not just acquire instrumental actions by observing others, as did early humans, but actively conform to the behavior and norms of the group, and even enforce conformity on others through teaching and social norm enforcement.

The combination of these changes in the two dimensions of human sociality created some totally new cultural realities. The transformative process was conventionalization, which has both a *coordinative* component, as individuals implicitly "agree" to do something in a consistent way (everyone wants to do it this way as long as everyone else does, too), and a *transmitive* component, as this way of doing things sets a precedent to be copied by others who want to coordinate as well. The result is what we may call *cultural practices*, in which individuals, in effect, coordinate with the entire cultural group via collectively known cultural conventions, norms, and institutions. In communication this means, of course, linguistic conventions, which serve their coordinative function because, and only because, they exist as "agreements" in the group's cultural common ground.

In terms of thinking, early humans imagined the world in order to manipulate it in thought via perspectival cognitive representations, socially recursive inferences, and social self-monitoring—which prepared them to coordinate with other specific individuals. But group-minded and linguistically competent modern humans had to be prepared to coordinate with anyone from the group, with some kind of generic other. This meant that modern human individuals came to imagine the world in order to manipulate it in thought via "objective" representations (anyone's perspective), reflective inferences connected by reasons (compelling to anyone), and normative self-governance so as to coordinate with the group's (anyone's) normative expectations. And these group-minded ways of operating and thinking were not present just in specific ad hoc collaborative interactions of the moment; rather, because of the way that modern humans became competent members of a cultural group during ontogeny, they created a permanent imprint in the human mind-set.

So once more let us look, first, at the new forms of collaboration evident in human cultural organization, then at the new forms of conventional linguistic communication for coordinating cultural life, and then at the resulting new forms of agent-neutral, normatively governed thinking that cultural life demanded.

## The Emergence of Culture

A number of animal species, from whales to capuchin monkeys, engage in one or another form of social transmission, requiring some form of social learning.

The most cultural of nonhuman animals are undoubtedly the great apes, especially chimpanzees and orangutans. Observations in the wild have documented for these two species a relatively large number of population-specific behaviors that persist in the group over time and that very likely involve social learning (Whiten et al., 1999; van Schaik et al., 2003). Experimental studies have also demonstrated some skills of social learning in these two species, for example, in learning to use novel tools, that very likely are at work in generating their cultural patterns in the wild (see Whiten, 2010, for a review).

But great ape culture is not human culture. Tomasello (2011) characterizes great ape culture as mainly "exploitive," as individuals socially learn from others who may not even know they are being watched. Modern human culture, in contrast, is fundamentally cooperative, as adults actively teach children, altruistically, and children actively conform to adults, as a way of fitting in cooperatively with the cultural group. The hypothesis is that this cooperative form of culture was made possible by the intermediate step of early humans' highly cooperative lifeways and how this transformed great ape social learning into truly cultural learning. Teaching borrows its basic structure from cooperative communication in which we inform others of things helpfully, and conformity is imitation fortified by the desire to coordinate with the normative expectations of the group. Modern humans did not start from scratch but started from early human cooperation. Human culture is early human cooperation writ large.

## Group Identification

The small-scale, ad hoc collaborative foraging characteristic of early humans was a stable adaptive strategy—for a while. In the hypothesis of Tomasello et al. (2012), it was destabilized by two, essentially demographic, factors.

The first factor was competition with other humans. This meant that a loose pool of collaborators had to turn into a proper social group in order to protect their way of life from invaders. A loose social grouping of early humans was under pressure to transform into a coherent collaborative group with joint goals aimed at group survival (each group member needing the others as collaborative partners for both foraging and fighting) and division-of-labor roles toward this end. As with early humans' smaller-scale collaborations, this meant that group members were motivated to help one another, as they were all now clearly interdependent with one another at all times: "we" must together compete with and protect ourselves from "them." Individuals thus began to understand themselves as members of a particular social group with a particular group identity—a culture—based on a we-intentionality encompassing the entire group.

The second factor was increasing population size. As human populations grew, they tended to split into smaller groupings, leading to so-called tribal organization in which a number of different social groupings were still a single supergroup or "culture." This meant that recognizing others from our own cultural group became far from trivial—and of course we needed to ensure that they could recognize us as well. Such recognition in both directions was important because only members of our cultural group can be counted on to share our skills and values and so be good and trustworthy collaborative partners. Contemporary humans have many diverse ways of marking group identity, but one can imagine that the original ways were mainly behavioral: people who talk like us, prepare food like us, and net fish in the conventional way—that is, those who share our cultural practices—are very likely members of our cultural group.

And so early humans' skills of imitation became modern humans' active conformity, both to coordinate activities more effectively with in-group strangers and to display group identity so that others will choose me as a knowledgeable and trustworthy partner. Teaching others to do things, perhaps especially one's children, became a good way to assist their functioning in the group and to ensure even more conformity in the process. Teaching and conformity then led to cumulative cultural evolution characterized by the "ratchet effect" (Tomasello et al., 1993; Tennie et al., 2009; Dean et al., 2012) in which modifications of a cultural practice stayed in the population rather faithfully until some individual invented some new and improved technique, which was then taught and conformed to until some still newer innovation ratcheted things up again. Tomasello (2011) argues that great ape societies do not display the ratchet effect or cumulative cultural evolution because their social learning is fundamentally exploitative and not cooperatively structured in the human way via teaching and conformity, which constitute the ratchet that prevents individuals from slipping backward.

The new sense of group identity characteristic of modern humans was thus extended not just in space to in-group strangers but also in time to ancestors and descendants in the group: this is the way "we" have always done things; it is part of who "we" are. As cultural practices were handed down across generations cooperatively—adults altruistically teaching and youngsters trusting and even conforming—the resulting cumulative effect was that the "we" became an enduring culture to which we (past, present, and future) are all committed (just as early humans were committed to their ongoing, small-scale collaborations). Human populations thus became more than a loosely structured pool of collaborators; they become self-identified cultures with their own "histories." Once again, precisely when this all happened is not crucial to our story, but the first clear signs of distinct human cultures appear with *Homo sapiens sapiens*, that is, modern humans, beginning at the earliest some 200,000 years ago.

That humans do indeed think of their group as a "we" of interdependent individuals—that humans identify with their group—is a well-established psychological fact. Most fundamentally, humans have a marked in-group/out-group psychology that is, in all likelihood, unique to the species. Much research shows that humans favor their in-group in all kinds of ways, and they care about their reputation more in their in-group than in any out-groups as well (Engelmann et al., in press). Moreover, they think of others from other groups not just as strangers, as do apes and as did early humans, but as members of specific out-groups with alien, often despised ways. Perhaps the most striking phenomenon of group identity is collective guilt, shame, and pride. Individuals feel guilty, ashamed, and/or proud when an individual of their group does something noteworthy in basically the same way that they would if they themselves had done the deed (Bennett and Sani, 2008). In the contemporary world, one sees such group identity and collective guilt, shame, and pride quite clearly in struggles over ethnic identity, linguistic identity, collective responsibility, and so forth—and even in such frivolous phenomena as fan support of sports teams. As far as we know, great apes do not have, and early humans did not have, this sense of group identity at all.

The proposal is thus that with increasing population sizes and competition among humans, the members of human groups began to think of themselves and their groupmates (known and unknown, present and past) as participants in one big, interdependent, collaborative activity aimed at surviving and thriving in competition with other human groups. Group members were identified most readily by specific cultural practices, and so teaching and conformity to the group's lifeways became a critical part of the process. These new forms of group-mindedness led to what we may call the *collectivization* of human social life, as embodied in group-wide cultural conventions, norms, and institutions—which transformed, one more time, the way that humans think.

## Conventional Cultural Practices

Group identification means that human groups each have their own set of conventional cultural practices. Conventional cultural practices are things that "we" do, that we all know in cultural common ground that we do, and that we all expect one another in cultural common ground to do in appropriate circumstances. Thus, in an open barter food market in which a conventionalized set of measurements for coordination is in place, if I show up with my honey in unconventional containers, no other traders will know what to do with me and my undetermined quantities of honey. With conventional cultural practices, deviations are not punished per se; they are simply left on the outside looking in. And there are some conventions that one cannot opt out of: one can wear this clothing, or that clothing, or nothing at all, but whatever one wears, it is a cultural choice that will either conform to or violate the expectations of others in the group.

Unlike the second-personal common ground that early human individuals created with one another as they engaged in collaborative activities, the common ground at this point is what Clark (1996) calls *cultural* common ground: things that we all in the group know that we all know even if we did not experience them together as individuals. Indeed, Chwe (2003) argues that the main function of the public events of a culture is to make sure that such things as the chief's coronation or his daughter's marriage ceremony become public knowledge: a part of the cultural common ground that everyone can count on everyone else knowing, which no one can plausibly deny knowing, and knowledge of which serves as a shibboleth of group membership. Interestingly, children as young as two years of age are already tuned into cultural common ground. Thus, Liebal et al. (2013) had two- and three-year-old children meet a novel adult (clearly from their group). This in-group stranger then asked them sincerely, "Who is that?" while they looked together at a Santa Claus toy and a toy the child had just made before the adult entered. Children answered by naming the newly created toy, as even children this young know that no one in the culture, not even someone they have never before met, needs to ask who is Santa Claus. (In a second condition they named Santa Claus if the stranger asked for the name of a toy she seemed to recognize.) Children in this same age range also expect that in-group strangers will know the conventional name of an object but not a novel, arbitrary fact about that same object (Diesendruck et al., 2010).

Some conventional cultural practices are the product of explicit agreement. But this is not how things got started; a social contract theory of the origins of social conventions would presuppose many of the things it needed to explain, such as advanced communication skills in which to make the agreement. Lewis (1969) thus proposed another way to get started. We begin with a coordination problem, say, what time to show up to go group fishing every day at our new camp. Let us say that by chance we depart on the first day at midday (just because that is when enough people for the task have congregated). Assuming no advanced communication skills, what do we do the next day? Following Schelling (1960), Lewis (1969) proposes that we search for anything to single out one time from all the others on this second day, and a natural way of doing that for humans seems to be "precedent"—we do what we did before (what worked for us before) and show up at midday again. And so we habituate, and new participants just imitate and conform to us. Anyone who does not conform simply does not participate.

But with the possibility of communication, we may also teach the convention to others and encourage them to conform and so to participate in the cultural practice. Importantly for our understanding of collective intentionality, when adults teach children how to perform cultural practices, the children take it not as communication about the current episodic event but, rather, as something general about the world, applying to things and/or events like these in general (i.e., "Fishing takes place at midday"). Thus, an adult might communicate to a child, without teaching, that there is a fish right there in the water. But when he goes into teaching mode the message is something more like, "These kind of fish are found in places like this," to facilitate the child's fishing skills in general (Csibra and Gergely, 2009). The implication of this pedagogical mode of cooperative communication is that there is a kind of objective reality that works on general principles (these kinds of fish in general, and these kinds of places in general), and the current situation is merely one instance of this objective reality. Teaching is implicitly backed by the collective and objective perspective on things developed by our cultural group.

Modern human children are thus learning from adults that there are certain ways that things should work. This "should" implicitly undergirding adult teaching prompts children, in ways that we do not fully understand, to objectify and reify the generic facts they are being taught into an objective reality—a generic perspective that is the ultimate adjudicator among differing perspectives on the world. This process has many implications for human thinking, but a prominent one is the understanding of false beliefs (which great apes clearly do not do; see Tomasello and Moll, in press, for a review). Thus, we previously invoked something like Davidson's notion of social triangulation to explain how it is that early humans came to understand that others have perspectives that differ from their own. But to get to an understanding of beliefs, including false beliefs, we must have some notion of a generalized perspective on an objective reality that is independent of any particular perspective. Something like this is needed to make the judgment not just that a belief is different from mine, but that it is wrong—since objective reality is the final arbiter. It is likely that young children begin to think in terms of multiple different perspectives on things from as soon as they participate in joint attention with its two perspectives during late infancy (Onishi and Baillargeon, 2005; Buttelmann et al., 2009), and we may hypothesize that this was the case for early humans as well. But it is not for several more years that children come to a full-blown understanding of beliefs, including false beliefs, because they (and so presumably all humans before modern humans) do not yet understand "objective reality."[1]

## Social Norms and Normative Self-Monitoring

In the small-scale collaborative interactions of early humans, individuals actively chose some collaborative partners and shunned others, and in some cases even rewarded and punished partners. But this was all done in second-personal mode, that is, as one individual evaluating another individual. What happened with modern group-minded humans was that these evaluations became conventionalized and so applied in agent-neutral, transpersonal mode, that is, applied by all to all (even by and to those not directly involved in an interaction) and with respect to objective, transpersonal standards. Although great apes retaliate for harm done to them, they do not punish other individuals for acts toward third parties (Riedl et al., 2012). In contrast, three-year-old children enforce social norms on others even when they are not personally involved or affected in any way, often using normative language about what one *should* or *should not* do in general (Rakoczy et al., 2008; see Schmidt and Tomasello, 2012, for a review).

   Social norms are thus mutual expectations in the cultural common ground of the group that people behave in certain ways, where the mutual expectations are not just statistical but, rather, socially normative, as in you are *expected* to do your part (or else!). The force of the expectations derives from the fact that individuals who do not conform to our group's way of doing things often create disruptions, which should not be tolerated, and indeed, if individuals behave too differently it signals that they are not one of us (or do not want to be one of us) and so cannot be trusted. Group-minded individuals thus view nonconformity in general as potentially harmful to group life in general. The result is that humans conform to social norms for instrumental reasons (to coordinate successfully), for prudential reasons (to avoid the group's opprobrium), and in order to benefit of the group's functioning since nonconformity potentially disrupts this functioning (a group-minded reason).

Like conventions in general, social norms operate not in second-personal mode but rather in agent-neutral, transpersonal, generic mode. First, and most basic, social norms are generic in that they imply an objective standard against which an individual's behavior is evaluated and judged. In early humans' social evaluations, individuals only knew who did things ineffectively or noncooperatively, but now the roles have specific agent-neutral standards (that can be taught as such). These objective standards come from the mutual understanding of how the different functions in particular conventionalized cultural practices are effected if everyone is to reap the anticipated benefit. Thus, if it is cultural common ground in the group that when collecting honey the person smoking out the bees must do so in this particular way, and that if she does not do it in this way we will all go home empty-handed, then her behavior may be evaluated relative to this objective standard for job performance.

Social norms are also generic in terms of their source. Social norms emanate not from an individual's personal preferences and evaluations but, rather, from the group's agreed-upon evaluations for these kinds of things. Thus, when an individual enforces a social norm, she is doing so, in effect, as an emissary of the group as a whole—knowing that the group will back her up. Group-minded individuals thus enforce social norms because their collective commitment to a social norm means that they commit not only to following it themselves but also to seeing that others do, too—for the benefit both of ourselves and of other group members with whom we are interdependent (Gilbert, 1983). The typical formulation of individuals enforcing social norms would be something like, "One cannot do it like that; one must do it like this," which is of course very similar to the generic mode used in teaching. (Indeed, norm enforcement and teaching may be two versions of the same phenomenon: enculturating individuals to our group's ways of doing and thinking about things.) The presupposition of norm enforcement is that this is the group's collective perspective and evaluation, or possibly even something generalized beyond that to some deity or some external normative facts of the universe: it is just the case in our world that this is the right way and that is the wrong way to do it.

And finally, social norms are also generic in their target: group disapproval is aimed in an agent-neutral way at in principle anyone, that is, anyone who identifies with our group's lifeways and so mutually knows and accepts in cultural common ground our social norms. (The common ground assumption exempts from the force of our social norms individuals from another social group, young children, and mentally incompetent persons [Schmidt et al., 2012].) This agent neutrality in application is nowhere more evident than in the fact that people apply social norms to themselves in acts of guilt and shame. Thus, if I take some honey that is needed by others, I will feel guilty under the force of the norm against stealing (perhaps undergirded by sympathy for the victim). More telling, if some illicit avocation of mine is made public, I will feel ashamed under the eyes of the group norm, even though I may not feel that it is wrong at all. Guilt and shame thus demonstrate with special clarity that the judgment being made is not my personal feeling about things (I wanted the honey and that avocation), but rather, it is the group's—which is especially clear in the case of shame, in which I may not even agree with the group. Nevertheless, I, as an emissary of the group, am sanctioning myself. Guilt and shame may thus have some second-personal bases in me feeling bad that I harmed another individual or that I did not conform to the expectations of valued others, but the full versions require in addition that I know I violated a collective norm. It is not just that my victim feels bad, or that I offended others' sense of decency but, more important, that the group—which includes me—disapproves.

Because I know that things work in this way, I self-monitor and self-regulate my actions via group norms so as to coordinate with the expectations of the group. In this normative self-monitoring, as we may call it, what one is often trying to protect is a public reputation, one's status as a cooperative member of the group (Boehm, 2012). That is to say, with modern human collaboration taking place on the level of the entire cultural collectivity, my behavior in various contexts might be known to some degree in the cultural common ground of the group as a whole (e.g., because of the pervasiveness of gossip, done best in a conventional language; see below). This means that early humans' concern with being judged is transformed by modern humans into a concern for one's public reputation and social status. And, critically, reputational status is more than just a sum of many social evaluations; it is nothing less than a Searlian status function (see next section) in which my public persona is a reified cultural product created by the collectivity, who can take it away in a second, as any scandalized modern politician can attest.

### Institutional Reality

In the limit, some conventional cultural practices turn into full-blown institutions. Obviously, the dividing line is fuzzy, but a basic prerequisite is that the cultural practice is not a solo activity but is in some sense collaborative, with well-defined, complementary roles. But the key feature distinguishing cultural institutions is that they comprise social norms that do not just regulate existing activities but, rather, create new cultural entities (the norms are not regulative but constitutive). For example, a human group might tend to make decisions about such things as where to travel next, how to set up defenses against potential predators, and so forth, by simply arguing among themselves. But if there are difficulties in making decisions, or infighting among several coalitions, then the group could institutionalize the process into some kind of governing council. Creating this council would give otherwise normal individuals abnormal status and powers. The council might then designate a chief, whom they would empower to do still other abnormal things, like banish people from the group. The council and chief are thus cultural creations, and their entitlements and obligations are bestowed upon them by the members of the group, who can, in theory, take them away and so turn the council members and chief back into everyday people again. The roles in institutions are explicitly agent neutral because, in theory if not in practice, anyone may play any role.

Searle (1995) has been most explicit about how this process works. First, obviously, is some kind of mutual agreement or joint acceptance among group members to designate, for example, an individual as chief. Second, there must be some kind of symbolizing capacity so as to enact Searle's well-known formula "X counts as Y in context C" (X counts as chief in the context of group decision making). Related to this, there should be some actual physical symbol—something like a leader's headdress or scepter or presidential seal—to help in marking the new status in a public way. The fact that institutions are public means that everyone knows that everyone knows about them and cannot plead ignorance in the face of overt symbolic marking. This is one reason why new institutions and officials are anointed with their new obligations and entitlements not just implicitly but explicitly and publicly. One could not do something bad to a chief wearing his official headdress, right after his official inauguration, and then plead ignorance of his status. Similarly with formally written rules and laws: their public nature essentially means that one cannot break them and expect to be excused by pleading ignorance.

Rakoczy and Tomasello (2007) argue that a simple model for understanding cultural institutions is rule games. Of course one may move a piece of wood shaped like a horse all around a checkered board in any way one likes. But if one wants to play chess, then one acknowledges that this horse-shaped piece is a "knight," and one moves a knight only in certain ways, and the other pieces in other ways, toward the goal of winning the game, where winning is defined by certain agreed-upon configurations of pieces. The pieces are given their statuses by the norms or rules, whose existence comes from, and only from, the explicit agreement of the players. Thus, we would argue that the ontogenetic cradle of such cultural status functions is young children's joint pretense when they, for example, designate together a stick to be a snake. In doing this they are engaging in the fundamental act that creates new statuses since, we would claim, this designation is a social, public agreement with one's play partner (see Wyman et al., 2009). Importantly, although the ability to pretend derives evolutionarily, as we have argued, from the way that early humans created pretend realities by pantomiming situations for others communicatively, the normative dimension comes only with the group-mindedness and collectivity characteristic of modern human cultures.

The most important point for current purposes is that there are in the modern human world social or institutional facts. These are objective facts about the world: Barack Obama is president of the United States, the piece of paper in my pocket is a €20 note, and one wins at chess by checkmating one's opponent. At the same time that they are objective, however, these facts are observer relative; that is, they are created by individuals in social groups, and so they may be just as easily dissolved (Searle, 1995). Barack Obama is president but only as long as we say so, euros are legal tender but only as long as we act so, and, in theory, the rules of chess may be changed at any time. What is absolutely extraordinary about social facts, then, is that they are both objectively real and socially created, speaking once again to the power of the objectification/reification process. Indeed, if one gives five-year-old children some objects with almost no instruction, they very quickly create their own rules for how to play with them, and they then apply these rules both to themselves and to new players as objective facts: "One must do this first," "It works like this," and so forth (Goeckeritz et al., unpublished manuscript). As in adult teaching and norm enforcement, the "must" here implies the guiding hand of an objective reality, independent of the perspective or wishes of any particular individual.

### Summary: Group-Mindedness and Objectivity

The social interactions of early humans were wholly second-personal. The social interactions of modern humans added on top of this a group-minded layer, starting with identification with one's group. Individuals in a particular cultural group know that everyone knows certain things, knows that everyone else knows them, and so on, in the cultural common ground of the group. There are collectively accepted perspectives on things (e.g., how we classify the animals of the forest, how we constitute our governing council) and collectively known standards for how particular roles in particular cultural practices should be performed—indeed, must be performed—if one is to be a member of the group. The group has its perspective and evaluations and I accept them; indeed, I myself help to constitute the group's perspective and evaluations, even if the target is myself.

Crucially, the generality involved in this new group-mindedness is not just schematicity. We are not talking here about an individual perspective somehow generalized or made large, or some kind of simple adding up of many perspectives. Rather, what we are talking about is a generalization from the existence of many perspectives into something like "any possible perspective," which means, essentially, "objective." This "any possible" or "objective" perspective combines with a normative stance to encourage the inference that such things as social norms and institutional arrangements are objective parts of an external reality. The generic nature of the communicative intention in both norm enforcement and pedagogy derives from the inherently generic group-mindedness and social normativity governing the way that "we" expect "us" to do things, which is then objectified into this is the way things are, or ought to be, in the world at large.

Early humans' dual-level cognitive model of jointness and individuality is thus scaled up by modern humans into a group-minded cognitive model of objectivity and individuality. Human group-mindedness thus reflects a profound shift in ways of both knowing and doing. Everything is genericized to fit anyone in the group in an agent-neutral manner, and this results in a kind of collective perspective on things, experienced as a sense of the "objectivity" of things, even those we have created. Thus is human joint intentionality "collectivized."

## The Emergence of Conventional Communication

In addition to conventionalizing their social lives in general into collective cultural practices, norms, and institutions, modern humans also conventionalized their natural gestures into collective linguistic conventions. Early humans' spontaneous, natural gestures of the moment were important for coordinating their many collaborative activities, but conventionalized gestures and vocalizations—collectively known by all of those who have grown up in our cultural group, past and present, but not by others—enabled much more decontextualized and flexible forms of communication and social coordination with all members of the cultural group, even those with whom one had never before interacted.

Given a naive view of the nature of linguistic communication, one might assume that the use of language obviates the need for thinking in the communicative coordination of intentional states: I encode my "meaning" in language, and you decode it—the way telegraph operators used to work with Morse code. But, in fact, this is not the way linguistic communication works (Sperber and Wilson, 1996). For example, a large proportion of words in everyday spoken discourse are pronouns (*he, she, it*), indexicals (*here, now*), or proper names (*John, Mary*), whose referents cannot be determined from a codebook but rather must be determined by accessing some kind of nonlinguistically constituted common conceptual ground. Moreover, our everyday discourse is liberally peppered with seemingly incoherent sequences, for example, Me: "Want to go to a movie tonight?" You: "I have a test in the morning." For me to interpret your answer as a "No," we must share in common ground an understanding that tests require studying beforehand, one cannot study and watch a movie at the same time, and so forth. This common ground then makes possible my abductive leap that you will not be coming with me to the movie.

And so the most basic thinking processes in linguistic communication are the same as in the account of pointing and pantomiming in chapter 3. In informative linguistic communication, I intend that you know something, and so I refer your attention or imagination to some situation (my referential act) in the hope that you will figure out what I intend you to know (my communicative intention). Then you, relying on our common ground (both personal and cultural), hypothesize abductively what my communicative intention might be, given that I want you to attend to this referential situation. Thus, I might come into your office and say, "The city of Leipzig is running a tennis camp for two months this summer." You comprehend my referential act perfectly well, but you have no idea why I am informing you of this situation. But then the abductive light bulb comes on: oh, if he intended to make a suggestion about what my children might do this summer (since we were talking about that last week), then referring my attention to this fact would make perfect sense. Anticipating this process, and in order to be an effective communicator, I attempt to simulate your potential abductive inference ahead of time and so formulate my referential act to guide it in the intended direction— just as in pointing and pantomiming. For example, I might anticipate that if I refer to "tennis camp" without referring to "this summer," she will think I mean a tennis camp for herself and not for her kids' summer activities—a process that is essentially me thinking about her thinking about my communicative act (which is itself an intention toward her intentional states).

Needless to say, communicative conventions add much more articulate semantic content to the communicative act than do pointing and pantomiming, perhaps making things easier, but they do not thereby obviate all of the simulating, inferring, and thinking that must be done for two language users to communicate about complex situations successfully. In addition, despite this commonality of process with the use of spontaneous gestures at some basic level, linguistic communication also provides some powerful new resources for human thinking. We may focus on four of these, each elaborated in its own section below: (1) communicative conventions as inherited conceptualizations, (2) linguistic constructions as complex representational formats, (3) discourse and reflective thinking, and (4) shared decision making and the giving of reasons.

### Communicative Conventions as Inherited Conceptualizations

Early humans' use of spontaneous iconic gestures as symbols to direct one another's attention and imagination to relevant situations now became, with modern humans, conventionalized in the group. This meant not only that interpreting a gesture depended on some personal common ground between communicative partners in the moment, as before, but also that it now depended on some cultural common ground about how we in this group expect others in this group to use and interpret this gesture (and expect others to expect us to, etc.). Thus, for instance, we all know in cultural common ground that when one wants to direct the attention or imagination of a partner to a snake-danger situation, one produces a wavy hand gesture in the direction of the potential danger. Such conventions are coordination devices in the sense that individuals want to use them only if everyone else uses them as well (Lewis, 1969; Clark, 1996). Communicative conventions thus come to be governed by constitutive norms in the sense that if I do not use them in the conventional way, I am just not in the game. As Wittgenstein (1955) argued so trenchantly, the criteria for conventional use are determined not by the individual but by the community of users. I can rebel, but to what effect?

This cultural dimension to communicative conventions—everyone in the group is expected in cultural common ground to know them and to conform to them—means that we can now consider human communicative acts to be fully *explicit*. Early humans made overt the way they were perspectivizing situations for others, for example, by pointing out relevant situations with a gesture, but the recipient could easily misunderstand, or even feign misunderstanding, and that would be the end of it. But now, if a modern human uses a communicative convention such as a snake-danger gesture, the partner cannot claim not to know it or, under normal circumstances, not to comprehend it. Since we all know the convention in cultural common ground, it is explicit, and so one must respond. The modern human was thus now not only under a second-personal pressure from the communicative partner to comprehend but also, in a sense, under normative pressure from the entire community: if you are one of us, then you know how to operate with this convention. Anyone not comprehending this communicative convention is just not one of us—which makes it culturally normative.

Iconic communicative conventions can quickly become noniconic. This happens reliably in the birth of sign languages, as isolated deaf individuals, who have been communicating their whole lives with hearing parents using spontaneous iconic gestures, come together and conventionalize some of these "home signs." This often leads to a kind of stylization or shortening of signs (see Senghas et al., 2004). Thus, the wavy hand for snake danger might become abbreviated to the point of almost no waving. Typically this is because the recipient can predict what is coming in the communicative situation; for example, if she is about to turn over a rock, as soon as someone sticks out his hand she can anticipate the snake-danger gesture. Children and other newcomers would then just imitate or conform to the abbreviated hand-out (with no waving) gesture to direct attention to snake-danger situations.[2] Powerful skills of imitation and conformity thus undermine iconicity in communication, as iconicity is not necessary in a group with cultural common ground about what gesture to use for communicating about certain situations conventionally. Communicative conventions can thus become "arbitrary."

The implications of this conventional-arbitrary way of doing things for the individual and her processes of thinking were, needless to say, momentous. For one thing, children were now born into a group of people using a set of communicative conventions that their ancestors had previously found useful in coordinating their referential acts, and everyone was expected to acquire and use exactly these conventions. Individuals thus did not have to invent their own ways of conceptualizing things; they just had to learn those of others, which embodied, as it were, the entire collective intelligence of the entire cultural group over much historical time. Individuals thus "inherited" myriad ways of conceptualizing and perspectivizing the world for others, which created the possibility of viewing one and the same situation or entity simultaneously under different construals as, for example, berry, fruit, food, or trading resource. The mode of construal was not due to reality, or even to the communicator's goals, but rather to the communicator's thinking about how best to construe a situation or entity so that a recipient would most effectively discern his communicative intention.

In addition to this fundamentally new form of conventional/normative and perspectival cognitive representation, communicating conventionally with arbitrary devices also creates, or at least facilitates, two other new processes of cognitive representation. The first is that the arbitrariness leads to a higher level of abstractness. Thus, when gestures are purely iconic, the level of abstraction is typically low and local. For example, with spontaneous iconic signs, opening a door is pantomimed in one way whereas opening a jar is pantomimed in another. This pattern is typical of the individually created home signs of deaf children, which are under pressure to remain iconic since there is no community of users with whom to conventionalize them. However, in a community, as the iconicity fades for new learners, and arbitrary conventions arise, there comes a more stylized depiction of *open* that is highly abstract and resembles no particular manner of opening. This abstractness is characteristic of many signs in conventional sign languages, and of course vocal languages as well. Conventionalization, after a drift to the arbitrary, breeds abstractness. One can imagine that acquiring large numbers of arbitrary communicative conventions could lead to some kind of general insight that most of the communicative signs we use have only arbitrary connections to their intended referents, and so, voilà, we can make up new ones as needed.

The second new process of cognitive representation created, or at least facilitated, by arbitrary communicative conventions also involves abstractness, but of a different type. Many of the most abstract conceptualizations in contemporary languages are single items for highly complex situations involving multiple agents doing things over time; for example, to define a term like *justice*, one would most naturally proceed with a kind of narrative: justice is when someone . . . and then someone. . . . It is difficult to imagine how to indicate for others complex situations and events such as *justice* in pantomime, except by acting out a kind of full narrative, and this is true even for more concrete narrative events like a celebration or a funeral, for which one would also have to pantomime whole sequences. But with arbitrary signs one may simply designate these complex situations with a single sign. This means that, in essence, arbitrary signs open up the novel possibility of symbolizing aspects of the relational, thematic, or narrative organization of human cognition—in addition to its straightforward categorical or schematized structure as symbolized by *tree* or *eat*—which expands the range and complexity of human thinking immensely. As argued in box 1 (chapter 3), the source of human conceptual access to this relational-thematic-narrative organization—which may now be designated with simple signs—is complex collaborative activities with joint goals and various constitutive roles. Markman and Stillwell (2001) refer to "role-based concepts" for the role slots (e.g., tracker in a hunt) and "schema-based concepts" for the overall activity (e.g., the hunting trip itself), and it is unlikely that any other organisms conceptualize this thematic dimension of experience.

Arbitrary communicative conventions—after there is a "critical mass"—also create two new processes of inference. First, because humans communicate for different purposes and at different levels of abstraction on different occasions, the individual in a community of conventional communicators inherits a large inventory of communicative conventions that relate to one another in complex ways. For example, one can imagine that in some contexts individuals would conventionalize a gesture or vocalization for indicating a gazelle, whereas on other occasions they would conventionalize a gesture or vocalization for indicating an animal in general (or maybe a potential animal prey of whatever species). Children in this culture would learn, in different contexts, both of these expressions. This now opens the possibility for not just causal inferences but formal inferences. If I indicate for you that a gazelle is over the hill, you may infer, based on your knowledge of things, that there is a potential animal prey over the hill, but you cannot make a similar inference in the opposite direction from animal to gazelle. Although in theory one could have spontaneous pantomimes at different levels of generality, it is only with collectively known conventionalized signs that communicators can be certain that recipients have the conventional means necessary to make such formal inferences—and so count on these inferences in formulating their communicative acts.

Second, arbitrary communicative conventions come to form a kind of "system" such that, precisely because of their arbitrariness, the referential range of one is constrained by the referential range of others in the same "semantic field" (Saussure, 1916). It is thus in our cultural common ground that I am making a choice between certain conventional expressions that we both know together I have available to me. For example, if I report to a friend that I saw his brother foraging with "a woman," the inference is that this was not his wife, even though his wife is a woman, too, because if it had been his wife I would have said "wife" and not "woman." Or, if I say that our child ate "some of the meat," the inference is that he did not eat all of it because—at least in the context where we are both quite hungry—if he had eaten all of it I would have said so. These kinds of pragmatic implicatures permeate discourse among contemporary language users based on the common cultural ground that we both have a certain inventory of conventional linguistic expressions from which we choose for communicative purposes. (Some of these inferences are recurrent and thus become what are often called, contentiously, conventional implicatures [Grice, 1975; Levinson, 2000].) Inferences of this type are not generated in the same way by spontaneous pantomimes or any other kinds of unconventionalized signs because in these cases it is not in the cultural common ground of the group that everyone knows all of the alternatives and so will be making inferences about a communicator's choice among them.

And so, with the advent of communicative conventions, we now have some new forms of conceptualization. Modern humans "inherit" a set of communicative conventions in their cultural common ground with others in the group, and the use of these conventions is normatively governed, in the sense that deviance from them puts one outside the cultural practice. The arbitrariness of communicative conventions means that they may be used to conceptualize situations and entities of almost unlimited abstractness, including role-relational, thematic, and narrative schemas. And with communicative conventions, we now have collectively known inferential connections among conceptualizations, both formal and pragmatic, which were not possible in the same way with natural gestures.

### Linguistic Constructions as Complex Representational Formats

If we imagine early modern humans with a small inventory of single-unit (holophrastic) communicative conventions, along with general cognitive abilities for creating novel mental combinations (possessed by all apes), we can easily imagine them creating multiunit linguistic combinations. And so, for example, perhaps there was a communicative convention for requesting eating by moving the hand to the open mouth. And perhaps there was an unrelated communicative convention for requesting going foraging for berries (miming a picking motion). It would not take a genius, in a situation in which someone offered something unpalatable to eat, to gesture eating followed by berries. Then, given an existing ability for schematization (possessed by apes and early humans, but now applied to conventions), one could imagine this individual generalizing the conventional eating gesture to other conventional food gestures, much like human toddlers do as they first say things like "more juice," followed soon by "more milk," "more berries," and so forth in a "more X" pattern (so-called item-based schemas; Tomasello, 2003a).

Linguistic constructions begin with such simple item-based schemas but then are elaborated and made more abstract through discourse interactions. The key aspect of the process for current purposes is the communicative pressure— the demand for sufficient information—coming from the recipient. This forces the communicator to make things explicit that he might otherwise have left implicit. Communicators stutter out a string of different utterances, pidgin style, and recipients have to fill in the gaps inferentially. But communicative breakdowns occur, and recipients demand more information about how to relate the bits to one another, and so communicators must be more explicit about their communicative intentions. This process—in combination with skills for integrating and automatizing sequences—turns things like "I spear antelope . . . he dead" into "I speared the antelope dead." When other, similar schemas are created (e.g., "I drank the gourd empty"), the result is a conventionalized linguistic construction, in this example the resultative construction (Langacker, 2000; Tomasello, 1998, 2003b, 2008). In the words of Givón (1995), today's syntax is yesterday's discourse.[3]

And so are born fully abstract linguistic constructions, which become gestalt-like symbolic conventions in their own right, with their own abstract communicative significance indicating different *types* of situations. For example, young English-speaking children learn early abstract constructions for (1) straightforward causal situations (e.g., the transitive construction: X VERBed Y); (2) causal situations viewed from the perspective of the affected object (e.g., the passive construction: Y got VERBed by X); (3) situations of object movement (the intransitive locative construction: X VERBed to/into/ onto Y); (4) situations of transfer of possession (e.g., the ditransitive construction: X VERBed Y a Z); (5) situations in which an agent acts with no affected object (e.g., the unergative intransitive construction: X smiled/cried/swam); (6) situations in which objects change states with no specification of the agent or cause (e.g., the unaccusative intransitive construction: X broke/ died); and so forth (Goldberg, 1995). Importantly, the communicative function of these abstract patterns is independent of the particular words used in them, as these schematic depictions illustrate.

In using a construction, communicators invite recipients to view or imagine situations from a particular perspective. Thus, in a conventional language there are various ways of designating subject, as perspectival topic, no matter who is performing or receiving the action. Thus, one and the same action may be referred to as "John broke the window," "The window got broken by John," "John's throwing of the rock broke the window," "The window was broken by John's throwing of the rock," "The rock broke the window," "The window was broken by the rock," and so on, depending of how the speaker wishes to conceptualize the situation for the listener on a particular occasion of use. Other constructions perspectivize the situation based on the communicator's judgment of the recipient's knowledge and expectations. For example, the English cleft construction, as in "It was John who broke the window," is used to indicate that John did the breaking when the recipient currently believes that someone else did; that is, it is used as a correction to a mistaken belief (e.g., You: "Bill broke the window." Me: "No, it was John who broke the window"). MacWhinney (1977) argues that these different construals derive from what communicators choose as their "starting point" or "perspective" for entering the event cognitively—which is conventionalized into the grammatical topic or subject.

From a cognitive point of view, abstract constructions give humans a new kind of abstract, syntagmatically organized conventional format for cognitive representation. These abstract constructions enable linguistic items to be used, and then reused, in a wide variety of different constructions, playing different roles on different occasions. Importantly, this flexibility in how items are used creates the need to explicitly mark the roles being played by different items. If I gesture or vocalize *man, tiger, eat*, it is important to know who is the agent and who is the patient of the eating activity. Modern-day languages have a variety of means for doing this, such as case marking and contrastive word order. Markers used to indicate participant roles may be seen as kind of second-order symbols, because they are about the role the participant is playing in the larger construction (Tomasello, 1992).[4] Importantly for current purposes, Croft (2001) argues that the linguistic items in an utterance gain their communicative functions not via their syntactic relations to other items but, rather, from the collaborative syntactic role they play in the utterance/construction as a whole. A linguistic construction may thus be seen, in a way, as a symbolic collaboration.

Abstract constructions are thus the main source of conventional linguistic productivity, and thus conceptual productivity in thinking. Individuals schematize and analogize to create abstract constructions, and they then can readily put new items into the slots of these constructions based on a more or less good fit with the communicative role of that slot. Indeed, using an item in a construction's slot coerces a construal that may be atypical for that item; for example, we quite often say such things as "he treed the cat," "he ate his pride," "he coughed his age," and so forth, which force atypical construals on some of the items. Such metaphorical or analogical thinking is testament to the fact that the construction itself has its own communicative function— from the top down, as it were—into which, up to some limit, items must be forced (Goldberg, 2006). In all, this system of conventionalized abstract constructions—and items that can be used and reused as needed in different constructions—enables the kind of creative conceptual combination by which humans may think, or at least try to think, of everything from flying toasters to colorless green ideas sleeping furiously.

All of this is about how the communicator specifies the referential situation that he and the recipient are communicating about during their ongoing second-personal communicative interaction. In addition, modern human communicators often use language as well to specify things about their communicative motive and their modal or epistemic relation to this referential material in their ongoing second-personal communicative interaction. Critically, this is something almost wholly new in the communicative process. Early humans made reference to things in the world in various ways, but their own relation to that referential material was left implicit—perhaps expressed unintentionally (procedurally) in a facial expression or vocalization, but not an intentional part of the communicative act under the communicator's voluntary control in decision making.

But now modern human communicators explicitly indicate their communicative motive in communicative conventions. Thus, most languages have different constructions for speech acts such as requestives and informatives (assertives). Philosophers of mind and language believe it to be very important that the "same" fact-like propositional content may be used in different constructions with different communicative motives, as in "She is going to the lake," "Is she going to the lake?," "Go to the lake!," "Oh, that she could go to the lake," and so forth. The idea is that this independence of the propositional content from any particular "illocutionary force" makes the propositional content into a kind of quasi-independent, fact-like entity, free of particular instantiations in particular linguistic utterances (e.g., Searle, 2001). As speech act function came to be expressed conventionally, in either linguistic items or constructions as a whole (as above), both the communicative motive and the propositional content were now conventionalized into the same representational format of words and constructions. In this totally new communicative move, the communicator's motive in the communicative interaction is now itself referred to and so conventionally conceptualized. This fact, among others, is what prompted Wittgenstein's (1955, #11) observation that a major difficulty in understanding how language works is that very different functions are all expressed in the same basic way in words and constructions: "Of course, what confuses us is the uniform appearance of words when we hear them spoken. . . . For their application is not presented to us so clearly."

In addition, communicators indicate through various linguistic devices their modal or epistemic "attitude" toward some propositional content in an utterance. Thus, a communicator might opine, modally, that "She must go to the lake" or that "She can go to the lake," or, epistemically, that "I believe she is going to the lake" or that "I doubt she is going to the lake." Again, the evolutionary raw material for the conventionalization of modal and epistemic attitudes is presumably facial expressions and prosody as expressed simultaneously with some utterance, for example, uncertainty, surprise, or indignation. But then these became conventionalized.[5] Communicators thus now encase propositional content also in a kind of "modal-epistemic envelope" (Givón, 1995)—again, with all items in the same conventionalized representational format of words and constructions—that further encourages us to conceptualize them as quasi-independent mental entities. In this case, the independence is not only from speaker motive but also from how speakers feel or think about them. This distinction between content and attitude is also foundational to the idea of some kind of timeless, objective, propositionally structured facts that are independent of how anyone thinks or feels about them and, therefore, also to the general idea of an independent, "objective" reality.

If we now combine all of the distinctions we have made—that is, those that the communicator actively controls (and for which he must choose from among alternatives)—we have the basic structure of a conventional linguistic utterance: the force-content distinction, within that the attitude-content distinction, and within that the topic-focus (subject-predicate) distinction, as shown in Figure 4.1.

Overall, then, we may say that linguistic constructions are conventionalized and automatized segments of discourse that organize human experience into abstract patterns of various sorts, as individuals conceptualize things for

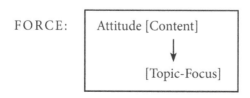

FORCE:    Attitude [Content]
              ↓
          [Topic-Focus]

FIGURE 4.1 The basic structure of a conventional linguistic utterance

others in communication. Constructions contain abstract roles—such as agent, recipient, location—marked via second-order symbols such as case markers, adpositions, or contrastive word order. The possibility of placing an almost unlimited inventory of linguistic items into these role slots is a major source of creative conceptual combination (see Clark's [1996] famous "The newspaper boy porched the newspaper"). Topic-focus (subject-predicate) organization within specific constructions serves to conceptualize situations from the perspective of one or the other of the roles. Communicative devices for indicating speaker motives, along with the modal-epistemic envelope, serve to partition out fact-like propositional content as indicating some kind of timeless, objective facts about an objective world, independent of how anyone thinks or feels about them. These are all aspects of human linguistic communication unique to the species (see box 3).

### Discourse and Reflective Thinking

Once we have linguistic communication, we have discourse. And what happens in discourse quite often is that the recipient responds to an utterance by signaling noncomprehension, requesting clarification, and so forth. The communicator then does his best to provide the needed information explicitly in further discourse. The key point for human thinking is that explicating conceptual content in conventional linguistic format (content that was only implicit in some original communicative act) makes this content ripe for self-reflection. That is to say, invoking again the analysis of Mead (1934; and to some degree that of Karmiloff-Smith, 1992), the collaborative nature of human communication means that the communicator can perceive and comprehend his own communicative act as if he were the recipient, which enables him to think about his own thinking, from an external perspective, as it were (see also Bermudez, 2003). Although the pointing and pantomiming of early humans enabled them to engage in some degree of reflection on their own overtly expressed thoughts, with modern humans and conventional linguistic communication, some new types of thoughts could now be expressed. Moreover, now the self-monitoring process came not just from the perspective of the recipient, but from the normative perspective of all users of the conventions. Three especially important examples are as follows.

First, one important piece of information often needing explication is the communicator's intentional states (or propositional attitudes). For example,

BOX 3.  The "Language" of Kanzi et al.

Over the past few decades, a handful of great apes have been raised by humans and taught some form of human-like communication. They end up doing very interesting things, but it is not clear in which ways they are human-like and in which ways they are not. With particular regard to linguistic constructions, there is no doubt that apes can combine their signs, sometimes creatively, but they do not seem to have anything resembling human constructions (even though they are perfectly capable of schematizing conceptual content in general). Why is this? To help answer this question, here are some examples of the kinds of utterances they produce, with either manual gestures or human-provided visual symbols (not resembling their referents):

BITE BALL—wanting to do this
GUM HURRY—wanting to have some
CHEESE EAT—wanting to
You(point) CHASE me(point)—requesting it from other

The first thing to note is that these are all requests, reflecting the fact that systematic studies have found that over 95% of the communicative acts produced by these individuals are some form of imperative (and the other 5% are questionable; Greenfield and Savage-Rumbaugh, 1990, 1991; Rivas, 2005). This is because no matter how they are trained by humans, great apes will not acquire a motive to simply inform others of things or share information with them (Tomasello, 2008). And in strictly imperative communication, there is little functional need for all the complexities of human linguistic communication (prototypically, no subject, no tense, etc.).

Nevertheless, many of the communicative acts produced by these individuals are clearly complex, structured by a kind of event-participant structure, reflecting a partitioning of situations into the participants involved in the relations or actions indicated. But despite this complexity, some key things relative to human linguistic communication are missing. Basically what is missing is all of those aspects of human grammar that conceptually structure constructions for others and their knowledge, expectations, and perspective. Over and above events and participants (and perhaps locations), the linguistic apes have learned items that indicate their own desire (e.g., although not apparent on the human gloss, their use of the item "hurry" indicates that they want it now). But what is missing is all of those aspects

(*continued*)

BOX 3 (*continued*)

of syntax that are geared at making the utterance comprehensible to the recipient—a key part of the cooperative motive. For example:

- They do not "ground" their acts of reference for the listener to help them identify the referent. That is to say, they do not have noun phrases with things like articles and adjectives that help to specify which ball or cheese is wanted, for example. Nor do they have any kind of markers of tense that would indicate which event, as indicated by when it occurred, they intend to indicate.
- They do not use second-order symbols such as case markers or word order to mark semantic roles and so to indicate who is doing what to whom in the utterance. Communicators do not need this information; it is provided to make sure that the listener understands the role of each participant in the larger situation or event being communicated about.
- They do not have constructions or other devices for indicating for listeners what is old versus new versus contrasting information. For example, if you adamantly expressed that Bill broke the window, I probably would correct you by using a cleft construction and say "No, it was FRED that broke the window." Apes do not have such constructions.
- They do not choose constructions based on perspective. For example, I might describe the same event either as "I broke the vase" or "The vase broke," based on your knowledge and expectations and my communicative intentions, whereas linguistic apes have not learned constructional alternatives of this type.
- They do not specifically indicate in their utterances their communicative motive (why should they, since it is always requestive) or anything of their epistemic or modal attitudes toward the referential situation.

The key theoretical point is that, beyond just supplying ordering preferences for utterances, human linguistic constructions are created with adaptations for the recipients' knowledge, expectations, and perspective in mind. And even very simple constructions like noun phrases require adaptations to the recipient's knowledge, expectations, and perspective. Humans also conventionalize expressions of motives and epistemic and modal attitudes in their constructions. Call all of this the pragmatic dimension of grammar, and call it uniquely human.

let us suppose that on my way back from a hunting trip I see gazelles drinking at watering hole no. 2, leading me to infer that their preferred watering hole no. 1 is currently dry (given the recent dry weather). Back at the home base, you inform me that you are headed to watering hole no. 1 to fetch water. I want to inform you that it very likely does not have water, but I do not want to just state as fact "It does not have water" since I am not certain. Presumably the first marker of speaker uncertainty used in such situations was some involuntary facial expression (see above). But then humans conventionalized ways of indicating doubt, for example, by saying something like, "Maybe it does not have water" or "I think it does not have water." Interestingly, young English- and German-speaking children first use words for thinking not to indicate a specific mental act of thinking but rather to express their uncertainty in the same way as *maybe* (thus, "I think it does not have water" means maybe it does not; Diessel and Tomasello, 2001). Only later do they make explicit reference to third-person mental happenings. And so one hypothesis is that it was the demands of discourse that led humans to begin talking explicitly about mental states, and they did not do this across-the-board initially, but only for their own epistemic attitudes toward propositional contents. Later, they came to refer to the mental states of anyone and everyone, including both others and the self, with the exact same set of communicative conventions. Once humans could make explicit reference to intentional states, they could think reflectively about them in some new ways.

A second set of cognitive processes often in need of explication is the communicator's logical inferring processes. These include most prominently those indicated by *and* and *or*, various kinds of negation (e.g., *not*), and implication (*if . . . then . . .* ). For example, in response to pressure from the recipient in argumentative discourse, the speaker requires terms such as these to make explicit his reasoning processes. And so, analogous to communicative pressure in normal discourse, "logical pressure" in argumentative discourse forces disputants to make explicit in language the logical operations that until that time were only procedural and not representational at all. One can imagine a first gestural/iconic step in which, for example, *or* is expressed in some kind of pantomiming in which one offers someone either *this* object (held out with one hand) or *that* object (held out with the other). An "*if . . . then . . .*" implication could be acted out in pantomiming such everyday social interactions as threats and warnings (if X . . . then Y). But, as always, symbolizing these logical operators in linguistic conventions would make them much more abstract and powerful and, once again, much more readily available for self-monitoring and self-reflection.

Third, speakers are often forced to make explicit some of the background assumptions and/or common ground to help the recipient to comprehend. For example, assume we are foraging together for honey, a cultural practice with which we are both very familiar from our cultural common ground. The knowledge we share about this practice—what kind of hive we are looking for, the height in the tree we should scan, the tools we will need, the container we will need for transport, and so forth—directs many of our activities. Thus, if you go off and start picking and weaving together leaves, I wait for you patiently as we both know that a vessel will be needed for transport. But this shared knowledge is all implicit in our (cultural) common ground. An early human might make this knowledge overt by pointing out to his partner the presence of some appropriate leaves. But now imagine that I, as modern human, express my intention that you notice the leaves' presence by means of some shared communicative conventions: "Look, there are some good leaves over there." This draws your attention to the leaves in a much more explicit way, but there is still room for misunderstanding (good for what?). So perhaps you look over at the leaves but draw a blank. Depending on my assessment of what you are not comprehending, I might say, "Its banyan leaves," or "We are going to need a vessel," or "We need banyan leaves to make the vessel," or whatever. I am making explicit for you the reason I am directing your attention to the leaves' presence (which I erroneously thought you could infer from our common ground), and, in the process, make explicit the bases for my own thinking for communicating. Once more, this makes it possible for me to reflect on my thoughts and their connections in a way that I could not when they were only an implicit part of our common ground.

And so with modern humans such things as intentional states, logical operations, and background assumptions could be expressed explicitly in a relatively abstract and normatively governed set of collectively known linguistic conventions. Because of the conventional and normative nature of language, new processes of reflection now took place not just as when apes monitor their own uncertainty in making a decision, and not as when early humans monitor recipient comprehension, but rather as an "objectively" and normatively thinking communicator evaluating his own linguistic conceptualization as if it were coming from some other "objectively" and normatively thinking person. The outcome is that modern humans engage not just in individual self-monitoring or second-personal social evaluation but, rather, in fully normative self-reflection.

### Shared Decision Making and the Giving of Reasons

We must single out, finally, a very special discourse context in human communication with world-changing implications for the process of human thinking: shared decision making. Prototypically, we may imagine as an example collaborative partners—or even a council of elders—attempting to choose a course of action, given that they know together in common ground that multiple courses of action are possible. Given their equal power in their interdependent situation, they cannot just tell the other or others what to do; rather, they must suggest a possible course of action and back it up with reasons.

Let us start with early humans. Because early human collaborators typically had much in common ground, they could point and pantomime in ways suggesting reasons implicitly. Thus, we may imagine two early humans following an antelope. They lose sight of the beast, and so pause in a clearing to make a joint decision about which way to go. One individual might in this context point to some tracks on the ground. They are relevant to the hunters because it is in their common ground that these are antelope tracks, possibly of the animal they were following. Also relevant is the direction of the tracks, which again is significant for the two of them because they know in common ground what this means for the antelope's likely travel direction. The point is that in directing his partner's attention to the tracks, the communicator's goal is that his partner travel with him in a certain direction. But he is not pointing in that direction; he is only pointing to the ground. The communicator's act is thus providing a kind of implicit reason for the recipient, which we may gloss as: see the tracks; given our common ground about what they mean for our prey's likely travel direction, they give us a reason for going in this direction. The recipient might counter by pointing in a different direction, where they spy the antelope's offspring next to some bushes, which is a better reason for traveling in this other direction. None of these reasons is explicit, of course, and so it does not constitute what we might call fully reasoned thinking. But it is a start.

With modern humans and their skills of conventional linguistic communication, we get to full-blooded reasoning, where "reasoning" means not just to think about something but to explicate in conventional form—for others or oneself—the reasons why one is thinking what one is thinking. This conflicts with the traditional view that human reasoning is a private affair. Most articulate on this point are Mercier and Sperber (2011) who recast the reasoning process in terms of communication and discourse, specifically argumentative discourse in which individuals make explicit to others their reasons for believing something to be the case. The basic idea is this: When a communicator informs a recipient of something, she wants to be believed, and often is (based on mutual assumptions of cooperation). But sometimes there is not enough trust on the recipient's part (for whatever reason), and so the communicator gives reasons for her informative statement. In reason-giving discourse of this kind, individuals are attempting to convince others. Many lines of evidence suggest that the main function of reasoning is to convince others, for example, people's tendency to look for supporting rather than disconfirming evidence (the confirmation bias). In this view, convincing others is good for individual fitness, and so humans evolved reasoning abilities not for getting at the truth but for convincing others of their views.

The proposal that human reasoning, including individual human reasoning, has a social-communicative origin is almost certainly correct. But Mercier and Sperber's account tends to background the cooperative processes involved, and so here is an alternative account that foregrounds these processes: The key social context is joint or collective decision making, as it occurred regularly in collaborative activities. Thus, on a hunting trip, perhaps you think we should hunt for antelopes in this direction, and I think we would be better off going in that direction. To make your case, you make your reasoning more explicit in our conventional language by, for instance, noting that there is a watering hole to the south. I counter, also in language, by making explicit my reasoning that at this time of day it is likely that lions will be at the watering hole and so no antelopes will be—and besides, here are some antelope tracks going to the north. You say these tracks look old, but I think that is because they were in the direct sunlight this morning and actually they are from around dawn or so. And on and on. The key point is that arguing in this way assumes a cooperative context. As Darwall (2006, p. 14) puts it: "It is only in certain contexts, say, when you and I are trying to work out what to believe together, that either of us has any standing to demand that one another reason logically."

Such cooperative argumentation, as we may call it, may be modeled in game theory as a battle of the sexes: our highest goals are collaborative—we will hunt together under all circumstances because otherwise there is zero hope of success—but within that cooperative framework we each argue our case. Critically, in this context, neither of us wants to convince the other if we are in fact wrong about the location of antelopes; each would rather lose the argument and eat tonight than win the argument and go hungry. And so a key dimension of our cooperativeness is that we both have agreed ahead of time, implicitly, that we will go in the direction for which there are the "best" reasons. That is what being reasonable is all about.

An appeal to "best" reasons invokes what Sellars (1963) calls "common standards of correctness and relevance, which relate what I do think to what anyone ought to think." Our cooperative argumentation in the context of joint or collective decision making is thus premised on a shared metric that we both use in determining which reasons are indeed "best." There have thus arisen social norms that govern cooperative argumentation in group decision making specifying, for example, that reasons based on direct observation trump reasons based on indirect evidence or hearsay. An even deeper, conceptual point is that to be in an argument in the first place means to accept as infrastructure certain "rules of the game," namely, the group's social norms for arguing cooperatively. This is the difference between a street fight and a boxing match. The early Greeks made explicit some of the most important of these norms of argumentation in Western culture, for example, the law of noncontradiction (a disputant cannot hold the same statement to be both true and false at the same time), and the law of identity (a disputant cannot change the identity of A during the course of the argument). Even before the Greeks, we can imagine that individuals who, for example, held the same statement to be both true and false at the same time were either ignored by others or else exhorted to argue rationally. The cooperative infrastructure was thus decisive in determining what it means to reason at all. The natural world itself may be totally "is"—the antelopes are where they are. However, the culturally embedded discourse processes by which we determine what that "is" in fact is—in the space of reasons, to use Sellars's evocative phrase—are fraught with ought.

Cooperative argumentation would thus have been the birthplace of "assertive" speech acts. Assertions go beyond the informative speech acts from which they derive in that the asserter commits himself to the truth of a statement (i.e., I commit not just to honesty but to the objective truth of the statement) and, crucially, to backing it up with reasons and justifications as necessary. Reasons and justifications are intended to make explicit to others the bases on which I believe something, which, because they share these bases, might give them reason to believe it as well (e.g., we all know and accept ahead of time that if there are lions at the watering hole, then there will be no antelopes). One may also reject an argument because it violates the norms of argumentation (e.g., you just contradicted yourself) or implies something that we both know is not true. Overall, this ability to connect thoughts to other thoughts (both those of others and one's own) by various inferential relations (prototypically by providing reasons and justifications) is key to human reason in general, and it leads to a kind of interconnection among all of an individual's potential thoughts in a kind of holistic "web of beliefs."

The capstone of all of this—recognized by all modern thinkers who take a sociocultural view of human thinking—is the internalization of these various interpersonal processes of making things explicit into individual rational thinking or reasoning. Making things explicit to facilitate the comprehension of a recipient leads the communicator to simulate, before actually producing an utterance, how his planned communicative act might be comprehended— perhaps in a kind of inner dialogue. Making things explicit to persuade someone in an argument leads the disputant to simulate ahead of time how a potential opponent might counter his argument, and so to make ready, in thought, an interconnected set of reasons and justifications—again, perhaps, in a kind of inner dialogue. As Brandom (1994, pp. 590–591) describes the process: "The conceptual contents employed in *monological* reasoning . . . are parasitic on and intelligible only in terms of the sort of content conferred by *dialogical* reasoning, in which the issue of what follows from what essentially involves assessments from the different social perspectives of scorekeeping interlocutors with different background commitments."

The norms of human reasoning are thus at least implicitly agreed upon in the community, and individuals provide reasons and justifications as ways of convincing "any rational person." Human reasoning, even when it is done internally with the self, is therefore shot through and through with a kind of collective normativity in which the individual regulates her actions and thinking based on the group's normative conventions and standards—what some have called "normative self-governance" (e.g., Korsgaard, 2009).

## Agent-Neutral Thinking

The second-personal thinking of early humans was aimed at solving coordination problems presented by direct collaborative and communicative interactions with specific others. Modern humans faced different kinds of coordination problems, namely, those involving unknown others, with whom one had little or no personal common ground. The solution on the behavioral level was the creation of group-wide, agent-neutral conventions, norms, and institutions, to which everyone expected everyone, in cultural common ground, to conform. To coordinate with others communicatively in such a world, human communication had to be conventional as well, based again not on personal but rather on cultural common ground. And to be a good communicative partner in conventional communication—especially, to be a cooperative participant in shared decision making—modern humans needed to express their reasons for thinking in certain ways explicitly in language and then simulate the cultural group's normative judgments of the intelligibility and rationality of those linguistic acts and reasons. Modern humans participate not only in joint intentionality with other individuals but also in collective intentionality with the entire cultural group.

### Representing "Objectively"

Early humans cognitively represented to themselves various situations and entities simultaneously from more than one perspective, and they then indicated or symbolized particular perspectives on those situations and entities for others in their deictic and iconic communicative acts. Modern humans then began collaborating and communicating with sometimes unfamiliar others structured by agent-neutral conventions, norms, and institutions, so that the cognitive models they were building and the perspectives they were simulating concerned not just particular others but, rather, some kind of generic other or, perhaps, the group at large. The linguistic conventions individuals were born into embodied the way that the group as a whole, from many years past, perspectivized and schematized experience, so that this way seemed inevitable. This new way of operating socially led to cognitive representations with three important new features.

CONVENTIONAL. For the first time in the history of life, modern human individuals "inherited" a culturally constructed representational system in the form of a conventional language, comprising a structured inventory of conceptualizations that forebears in the culture had previously found useful in communicating with others. The uses of linguistic conventions were shared within the cultural common ground of the group, and this meant, given human group-mindedness and conformity, that they became normatively grounded in "community standards" governing their proper use. This made it seem, especially to language-acquiring children, that the way that one's linguistic conventions carved up the world was somehow natural.

In addition, the arbitrariness of linguistic conventions created, or at least facilitated, the ability to operate with highly abstract conceptualizations such as *justice* or *blackmail* that schematize not a taxonomic class but, rather, a thematically or narratively defined entity. The arbitrariness of linguistic symbols also led to more abstract conceptualizations of relatively concrete terms, such as *open* or *break*, across different particular situations. Most important, because of their conventional nature, linguistic conventions and their interrelations made possible conceptualizations with explicitly contrasting aspectual shapes— *gazelle, animal, dinner*—known to everyone in the cultural common ground of the group. These contrasting aspectual shapes created a gap for the individual between the world as she conceptualized it for her own instrumental actions and the world as conceptualized in various contrastive ways in her conventional language, a gap that humans have been pondering from the early Greeks to Benjamin Lee Whorf.

PROPOSITIONAL. Modern humans began to use linguistic conventions together in combination in patterned ways that led to the creation of abstract linguistic constructions as kinds of linguistic gestalts. Many linguistic constructions conceptualize whole propositions, and they do this with internal constituents that are marked with second-order symbols as playing specified roles in the construction. Proposition-level linguistic constructions are perspectival (e.g., active vs. passive), and one of the elements (subject) provides a perspectival entry point into the conceptualized situation. The abstractness of linguistic constructions makes possible especially productive conceptual combinations, so that we may represent to ourselves all kinds of imaginary entities and situations, from a happy sun to a man in the moon. Linguistic constructions thus create the possibility of various kinds of metaphorical representations in which structural analogies provide a new framework for thinking, from one idea "undermining" another to activities "eating up" my free time. Making explicit in linguistic constructions various kinds of communicative motives and attitudes contributed to an objectified view of experience, as it suggested a fact-like propositional content independent of the desires or attitudes of any particular individual.

With their linguistic constructions, modern humans also began to make assertions—to whose objective truth they were committed—that could be either about particular episodic events or else, importantly, about generic events or facts of a type. They did this especially in norm enforcement ("One does not do that in public") and teaching ("It works this way"). This genericness presumably originated from the normative "group voice" that lay behind the assertive expression and gave it an objectivity that transcended the individual.

"OBJECTIVE." Early humans lived in a world of different individual perspectives. Modern humans live in this world too, but in addition, in the context of group-minded cultures, there arose a kind of public world comprising collectively created entities such as conventions, norms, and institutions from marriage to money to governments. These entities existed before the individual arrived on the scene, and they existed independent of the thoughts and wishes of any single individual, giving them the same kind of "always already there" status as the physical world. In addition, these collective entities had preestablished roles into which, in theory, any agent could seamlessly fit; and, indeed, in some cases these roles created new realities, such as presidents and money, whose very real deontic powers were readily observable. Operating in this public world required that individuals be able to take a kind of agent-neutral perspective on things, a kind privileged, "transcendental" perspective that constituted the world "objectively" and that then justified personal judgments of true and false, right and wrong.

As modern human individuals were building their cognitive models of the world, the use of simple causal and intentional relations was not enough. To explain such things as chiefs and marriage, not to mention language and culture, they needed some understanding of things created by collective agreement and maintained by collective normative judgment. Said another way, they needed some new conceptualizations of collective realities that transcended the thoughts and attitudes of single individuals, even multiple individuals. Constructing such models would lead naturally to judgments such as *real*, *true*, and *right* that come not from the individual herself but rather from her appropriation of the transpersonal, "objective" perspective engendered by her cultural world. Linguistic representations—especially assertions with a distinction between the second-personal attitudes of the communicator and some timeless, generic propositional content (e.g., "I think its raining")—only added additional force to these objectifying and reifying tendencies. Modern humans thus "collectivized" early humans' ways of life, and so "objectified" their cognitive models of the world.

### Reasoning Reflectively

The inferences of the common ancestor to humans and great apes were simple causal and intentional inferences. The inferences of early humans were recursively structured, enabling them to produce and interpret cooperative communicative acts comprising nothing but a protruding finger. But now, the linguistic communication of modern humans opened up whole new vistas of inference and reasoning. We now have such things as formal and pragmatic inferences, and external communicative vehicles can be reflected upon by the communicator from an objective and normative perspective. And the giving of reasons and justifications to others—and to the self in internal reasoning—now serves to connect up an individual's various conceptualizations into a single inferential web.

LINGUISTIC INFERENCES. The hierarchical relationship between the referents of different linguistic conventions is part of the conventionalization process. Thus, everyone knows collectively that everyone uses *gazelle* only for a particular type of animal, and *animal* for all kinds of animals, of which gazelle is one type, and so we now have the possibility of formal inferences: if we know that a gazelle is over the hill, then we know that an animal is over the hill (but not the reverse). Much of the early development of formal logic was built on inferences of this type, and in contemporary conceptual role semantics, inferences of this type play an important role as well. Also crucial is the fact that we all know collectively that we all know the linguistic options available to a communicator, which leads to the kind of pragmatic inferences that Grice (1975) made famous: if I refer to someone as an "acquaintance," that almost certainly means that we are not friends—because if we were friends I would have used the word *friend*. These implicatures and corresponding inferences are possible because, and only because, the options available are part of the group's cultural common ground, so that we can wonder together why I made the choice that I did. Conventional linguistic communication thus makes possible powerful new kinds of inferences.

In addition, linguistic communication, and the arbitrary nature of linguistic conventions, enabled modern humans to express explicitly in language some conceptualizations that could not be expressed easily, if at all, in the natural gestures of early humans, for example, intentional states and logical operations. Based on the hypothesis that one can reflect on one's thinking only as it is expressed in external behavior directed at another—because only then can one play the other's role and attempt to comprehend it from her perspective—linguistic communication now made available to modern humans many new conceptualizations about which they could think reflectively. Importantly, as modern humans thought about their own thinking reflectively, at least in some situations, they did not do so merely from their own perspective, or even that of the other, but from a more "objective" perspective.

REFLECTIVE INFERENCES. A special discourse situation is cooperative argumentation, in which we attempt to come to a group decision about either action or beliefs. We do this not only by making assertions, committed to the truth, but also by backing up those assertions with reasons and justifications, which means making connections to things that are collectively agreed to be true and reliable. The outcome of this process is that the various conceptualizations and propositionally structured thoughts of modern humans as expressed in language become ever more inferentially interconnected in a vast "web of beliefs," such that each element in the web gains significance from its inferential relations with others. This interconnectedness is a key component in being a fully rational creature who "knows his way about" an entire conceptual system in which propositionally structured thoughts provide reasons and justifications for one another (i.e., they can be used as premises and conclusions for one another in argumentation; Brandom, 2009).

In addition, as modern humans began engaging in such cooperative argumentation, implicitly accepted norms emerged. These norms operated such that individuals who contradicted themselves from one assertion to the next, or changed the meaning of their terms in the middle of an argument, or held a single assertion to be both true and false, were basically ignored or excluded from the group decision-making process. The process of cooperative argumentation, then, was the special language game within which human norms of rationality came to govern all those who wanted a voice in collective decisions (such as political, judicial, and epistemic decisions).

All of this may be internalized. Internalization means simply that one directs a communicative act, as communicator, to oneself, as recipient, including holding the "other" to "objective" normative criteria of intelligibility, cooperative participation, and so on. The resulting internal dialogue is one especially salient type of human thinking (Vygotsky, 1978). When the communicative context is cooperative argumentation, what now get internalized are whole lines of argumentation and justifications for arguments. Now, an individual can give to himself a normatively justified reason for why he is thinking what he is thinking, and so his conceptualizations become defined, in large part, by their normatively sanctioned inferential relations with other conceptualizations. The resulting web of beliefs, and humans' ability to navigate this web facilely, is foundational for the ability to engage in individual reasoning.

At this point, then, the inferring of modern humans is not just imagining causal and intentional sequences, as in apes, or even just perspectivizing and recursivizing them, as in early humans; rather the inferring of modern humans now includes new kind of inferences made possible by a conventional language, and new forms of reflecting on their own thinking, including in a kind of inner dialogue. When these processes operate in the special context of cooperative argumentation, the result is something we might call reasoning. Modern humans thus are on occasion able to engage in a kind of reasoned, or reflective, inferring in the context of the normative standards of the cultural group.

## Normative Self-Monitoring

Early humans engaged in what we have called cooperative self-monitoring—regulating their collaborative activities by the evaluative reactions of specific partners; and communicative self-monitoring—regulating their communicative acts by the anticipated interpretations of specific partners. Scaling up these processes to the cultural way of life characteristic of modern humans means that individuals now regulate their behavioral decision making instead via the collectively known and collectively accepted norms of the cultural group. Modern humans thus came to feel not only a second-personal pressure in their decision making but also, on top of this, as it were, a group-level normative pressure to conform to the group. Thus, I do not renege on my commitments, first of all, because I do not want to disappoint my partner, and second of all, because "we" in this group do not treat others like that. This more generalized normativity thus ends up back at group identity: if I want to be a member of this group, I must behave as they do, that is, follow the norms to which we all together (including me) have committed ourselves.

Modern human thinking and reasoning become normatively structured and governed in multiple ways. When one communicates with others via communicative conventions, one needs to do it in the way that they do it to participate successfully. In addition, in the context of group decision making and cooperative argumentation, one must agree to certain norms of argumentation. Others in the group decision making have a stake in me participating in useful ways, and so everyone has a stake in everyone else making true assertions, following the norms of inference and argumentation, justifying by connecting to propositions and arguments already collectively accepted, and so forth. Internalized, this communicative process becomes individual reason.

NORMATIVE SELF-GOVERNANCE. Normative self-governance results from an internalization of the processes of collective normativity, as the individual self-monitors and, indeed, self-regulates her actions by taking into account the social norms of the group, both cooperative and communicative. Modern humans communicate with themselves and so reflect on and evaluate their own thinking with group-held normative standards. This reflection means that humans know what they are thinking and can provide to themselves normatively sanctioned justifications and reasons for thinking in this way— thus connecting up their many and diverse thoughts in an intricate inferential web that is governed, to some extent, by "community standards." The thinking subject also uses this reflection in exercising executive control over her own thinking and reasoning. Korsgaard (2009), in particular, has emphasized that humans not only have goals and make decisions and reason in particular ways but also attempt to assess ahead of time whether those are good goals to pursue or good decisions to make or good reasons to have—a clearly extra layer of reflection and evaluation. And the normative judgment here is not simply mine alone, nor that of a specific other partner, but rather a judgment about whether that would be a good goal or decision or line of reasoning for any rational person, that is, for anyone from our group who does things the way that we do them.

Modern humans thus operate with the social norms of the group as internalized guides to both action and thinking. This means that in their collaborative interactions modern humans conform to the collectively accepted ways of doing things, based on norms of cooperation, and in their communicative interactions they conform to the collectively accepted ways of using language and also linguistically formulated arguments, based on the group's norms of reason.

## Objectivity: The View from Nowhere

Unlike other great apes, who all live in the general vicinity of the equator, modern humans have migrated all over the globe. They have done this not as individuals but as cultural groups; in none of their local habitats could a modern human individual survive for very long on his own. Instead, in each specific environment, modern human cultural groups have developed collectively a set of specialized and cognitively complex cultural practices to accommodate the local conditions, from seal hunting and igloo building to tuber gathering and bow-and-arrow making—not to mention science and mathematics. What we have attempted to do here is to specify the skills of cognition and thinking that enable modern human individuals to coordinate with those around them, both collaboratively and communicatively, in their efforts to adapt together to the novel exigencies they encounter in their specific corner of the world.

An image for the advent of modern humans is this: Earlier humans are living quite nicely by collaborating and communicating with others in various ways for a variety of cooperative purposes. Then, in the face of some serious demographic challenges, a great wave of group-mindedness and conformity washes over everyone. Humans who were spontaneously coordinating with partners to hunt or gather their daily meals then began to develop for foraging a number of conventionalized cultural practices. Humans who were spontaneously communicating with their partners in coordinating their complex collaborative activities using ad hoc gestures then began developing skills of conventional linguistic communication. And humans who were spontaneously exhorting or admonishing one another second-personally in various cooperative directions then began to develop collectively known and applied social norms of morality and rationality. Early humans lived together and interacted with others jointly; modern humans lived together and interacted with others collectively.

One effect of this great wave of group-mindedness and conformity was cultural group selection accompanied by cumulative cultural evolution. Cultural group selection takes place when individuals conform within their group—and differentiate themselves from other groups—to the extent that the group itself becomes a unit of natural selection (Richerson and Boyd, 2006). In this way, successful cultural adjustments to local conditions stay around, and unsuccessful attempts die out. Cumulative cultural evolution takes place when the inventions in a cultural group are passed on with such fidelity that they remain stable in the group until a new and improved invention comes along (the so-called ratchet effect; Tomasello et al., 1993). Modern humans had a stronger ratchet than early humans and apes because they had—in addition to powerful skills of imitation—proclivities both to teach things to others and also to conform to others when they themselves were being taught. And so it is with this wave of group-mindedness and conformity that we get the possibility of cultural groups creating and constantly improving their own cognitive artifacts—from procedures for whale hunting to procedures for solving differential equations—that help them both to adapt to local conditions and to mark themselves as distinct from other cultural groups.

The subterranean effect of this wave of group-mindedness and conformity, as it were, was new and culturally collective forms of cognitive representation, inference, and self-monitoring for use in thinking. Modern humans began representing the world "objectively," reflecting a kind of generic, agent-neutral perspective possible by any rational person. Further, humans' new skills of conventional linguistic communication enabled them to talk about many things that they previously could not (e.g., mental states and logical operations), and this enabled reflective inferences—thinking about one's own thinking—with much greater depth and breadth. In the context of cooperative argumentation, modern humans made explicit the reasons for their assertions, thus connecting them in an inferential web to their other knowledge, and then this social practice of reason-giving was internalized into fully reflective reason. And the self-monitoring of modern humans for the first time reflected not just their expectations about the second-personal evaluations of specific others but, rather, their expectations about the normative evaluations of "us" as a cultural group. Given all of these new ways of behaving and thinking, the crack in the human experiential egg now became a veritable chasm: the individual no longer contrasted her own perspective with that of a specific other— the view from here and there; rather, she contrasted her own perspective with some kind of generic perspective of anyone and everyone about things that were objectively real, true, and right from any perspective whatsoever—a perspectiveless view from nowhere.

And so, if from a moral point of view, cooperation always entails some kind of effacing of one's own interests in deference to those of others or the group, then, from a cognitive point of view, cooperative thinking always entails some kind of effacing of one's own perspective in deference to the more "objective" perspective of others or the group (Piaget, 1928). Thus, in cooperative communication I must always honor the perspective of my recipient, and in cooperative argumentation I must be committed to accept the reasons and arguments of others if they are better than my own—by the yardstick of our agreed upon normative criteria of rationality, which include our agreed upon objective reality—and so to abandon mine for theirs. In the words of Nagel (1986. p. 4): "Objectivity is a method of understanding . . . To acquire a more objective understanding of some aspect of life or the world, we step back from our initial view of it and form a new conception which has that view and its relation to the world as its object. . . . The process can be repeated, using a still more objective conception." In this formulation, "objectivity" is the result of being able to think of things from ever wider perspectives and also recursively, as one embeds one's perspective within another, more encompassing perspective. In the current view, more encompassing means simply from the perspective of an ever wider, more transpersonally constituted generic individual or social group—the view from anyone.

The monumental second step on the way to modern humans thus took the already cooperativized and perspectival thinking of early humans and collectivized and objectified it. Whereas early humans internalized and referenced the perspective of what Mead (1934) calls the "significant other", modern humans internalized and referenced the perspective of the group as a whole, or any group member, Mead's "generalized other." Human thinking at this point is no longer a solely individual process, or even a second-personal social process; rather, it is an internalized dialogue between "what I do think" and "what anyone ought to think" (Sellars, 1963). Human thinking has now become collective, objective, reflective, and normative; that is to say, it has now become full-blown human reasoning.

# 5

## Human Thinking as Cooperation

The internalization of socially rooted and historically developed
activities is the distinguishing feature of human psychology.

—LEV VYGOTSKY, *MIND IN SOCIETY*

Human cognition and thinking are much more complex than the cognition
and thinking of other primates. Human social interaction and organization are
much more complex than the social interaction and organization of other primates as well. It is highly unlikely, we would argue, that this is a coincidence.

Complex human cognition is of course responsible for complex human
societies in the sense that human societies would fall apart if human-like cognition were not available to support them. But this cognition-to-society
causal link is not a plausible direction for an account of evolutionary origins.
For that direction of effect, there would need to be some other behavioral
domain in which powerful cognitive skills were selected, and then those
skills were somehow extended to solving social problems. But it is not clear
what other behavioral domain that might be, given that we are trying to explain the many particularities of cognitive skills supporting humans' unique
forms of collaboration and communication, including in the end such things
as cultural conventions, norms, and institutions. It seems highly unlikely
that cognitive skills adapted for, say, individual tool use or the tracking of
prey could be exapted in this way for such complex cooperative enterprises.

And so, in the current view, the most plausible evolutionary scenario is
that new ecological pressures (e.g., the disappearance of individually obtain-
able foods and then increased population sizes and competition from other
groups) acted directly on human social interaction and organization, leading
to the evolution of more cooperative human lifeways (e.g., collaboration for
foraging and then cultural organization for group coordination and defense).
Coordinating these newly collaborative and cultural lifeways communica-

tively required new skills and motivations for *co*-operating with others, first via joint intentionality, and then via collective intentionality. Thinking for *co*-operating. This, in broadest possible outline, is the shared intentionality hypothesis.

But our evolutionary story has taken many more detailed twists and turns as we have attempted to account, in detail, for the many different aspects of uniquely human thinking as they relate, in detail, to the many different aspects of uniquely human collaboration and communication. Because there are no other contemporary evolutionary stories with exactly this focus, we have thus far made scant reference to other theories. But there are a number of other contemporary accounts of the evolution of uniquely human cognition and/or uniquely human sociality in general, and a broad survey of these will help to better situate the shared intentionality hypothesis within the current theoretical landscape.

## Theories of Human Cognitive Evolution

When asked what makes human cognition and thinking unique, a kind of default answer for many cognitive scientists would be something like "general intelligence." Since humans have evolved very large brains (roughly three times larger than those of other great apes), and since larger brains have more computing power, the idea is that humans are able to engage in all kinds of cognitive processing, including thinking, in bigger, better, and faster ways. But even if this description is in some sense true, the question remains of how it came about evolutionarily. It is implausible in the extreme to just say that being smart is more adaptive than being dumb and so humans became smarter. This is a just-so story of the most egregious kind. Being able to fly and walk would be better than being able to walk only, so why did humans not come to fly as well? The point is that a plausible evolutionary account must be built on an adaptive scenario involving a specific set of circumstances in which a specific set of cognitive skills provided a specific set of advantages to the individuals who possessed them.

In the case of general intelligence, if this is a useful construct at all, recent data suggest that the more specific story would almost certainly be some variant of our social account. Thus, Herrmann et al. (2007, 2010) administered a comprehensive battery of cognitive tests—assessing skills for dealing both with the physical world and with the social world—to large numbers of two of human's closest primate relatives, chimpanzees and orangutans, and to 2.5-year-old human children. If the difference between human and ape cognition were based on general intelligence, then the children in this study should have differed from the apes uniformly across all the different tasks. But this was not the case. The finding was that the children and apes had very similar cognitive skills for dealing with the physical world, but the children—old enough to use some language but still years away from reading, counting, or going to school—already had more sophisticated cognitive skills than either ape species for dealing with the social world. The hypothesis was thus that human adults are cleverer than other apes at almost everything not because they possess an adaptation for greater general intelligence but, rather, because they grew up as children using their special skills of social cognition to cooperate, communicate, and socially learn all kinds of new things from others in their culture, including the use of all of their various artifacts and symbols (Herrmann and Tomasello, 2012).

A similar but different argument applies to theories that propose narrower, but still domain general, cognitive processes to distinguish humans from other primates. The most systematic such attempt is that of Penn et al. (2008). They claim that what distinguishes human from nonhuman primate cognition is humans' ability to understand and reason with various kinds of higher-order relations. In addition to several empirical disputes about the data they cite for great apes, the overall problem is that this theory would also predict across-the-board differences in how well humans and other great apes deal with various kinds of problems in different domains of activity. But, again, the Herrmann et al. (2007, 2010) results are not consistent with this account. Furthermore, Penn et al. have no evolutionary story specifying the adaptive context(s) that could account for humans' special skills with relational conceptualizations. The discussion in box 1 (chapter 3) proposed an alternative account, namely, that humans' especially sophisticated relational thinking derives from comprehension of the individual roles involved in various types of joint and collective intentionality. Thus, this special form of relational thinking is just one outcome of the process of adapting cognitively to new forms of social engagement. And something similar may be said about Corbalis's (2011) proposal that the key to human cognitive uniqueness is recursion, especially as manifested in language, "mental time travel," and theory of mind. Recursion also plays a key role in the current account, but again, we would claim that it is not the whole story. Rather, it is an outcome of the process by which humans came to collaborate and communicate with others in special ways; specifically in this case, it is the special way that individuals had to make inferences to participate in cooperative (ostensive-inferential) communication.

A second set of hypotheses to explain human cognitive uniqueness invoke language and/or culture. In the case of language, some theorists have pointed to the unique kinds of computational processes that language enables, namely, various kinds of combinatorial/syntactic productivity, including recursion (see Bickerton, 2009, for a recent version of this view). More philosophically minded theorists have focused on the role of language in reasoning, that is, on the way that humans make assertions aimed at truth, and then attempt to justify them to others with articulated reasons (as in science and mathematics and, perhaps, courts of law and political disputes), and this is only possible in the medium of a language of some kind (see, e.g., Brandom, 1994). Of course, no one can dispute the crucial role of language in human thinking—and language is a key part of our proposed second step in human cognitive evolution—but, in the current view, it plays its role only fairly late in the process. Indeed, we have argued previously that human language was made possible by a number of earlier adaptations for joint intentionality (e.g., joint goals, common conceptual ground, recursive inferences), and that its eventual emergence was part of a larger process in which many human activities were conventionalized and normativized (Tomasello, 2008). In our view, saying that only humans have language is like saying that only humans build skyscrapers, when the fact is that only humans, among primates, build any kind of stable shelters at all. Language is the capstone of uniquely human cognition and thinking, not its foundation.

Somewhat relatedly, many social and cognitive anthropologists have insisted that what is most remarkable about human cognition, compared with that of other primates, is its variability across different human populations, which attests to its grounding in processes of culture (e.g., Shore, 1995; Chase, 2006). More radically, various postmodern theorists have claimed that basically all of human experience takes place within the discursive practices of a human culture, and so uniquely human thinking is only imaginable within this cultural framework (e.g., Geertz, 1973). Again, these claims for the key role of culture are all, in some general sense, true. But again, if our question is evolutionary origins, they are not sufficient. Human thinking became unique even before the flourishing of human cultural variability, specifically, in the evolution of species-wide skills of collaboration, cooperative communication, and joint intentionality more generally (and it can be seen today in the species-unique skills of prelinguistic human children). These skills then enabled the evolution and development of culture at a subsequent time. This analysis also applies to the account of Richerson and Boyd (2006), who have argued for the crucial role of cultural group selection in most things uniquely human. Again, the second (cultural) step of our story invoked this process as well, but again, there are numerous prerequisite and concomitant uniquely human capacities that make cultures, and consequently cultural group selection, possible in the first place (e.g., conformity, conventionalization, and normativization). Since a culture comprises conventionalized ways of doing things, for modern human cultures to be the way that they are, some things must have already been complex and fundamentally cooperative "naturally," before they were conventionalized.

And so we agree with almost everyone that language and culture were necessary for the evolutionary emergence of modern human cognition and thinking. We have just argued that they were made possible by other uniquely human social and cognitive processes—namely, those associated with joint and collective intentionality more generally—that emerged earlier or concomitantly in human evolution. A full account must therefore acknowledge the role of these earlier and/or concomitant processes, and indeed, our own view is that an understanding of how language and culture work as modes of social engagement and interaction at all requires a full explication of the underlying processes of joint and collective intentionality involved (Tomasello, 1999, 2008).

The third and final set of hypotheses comes from evolutionary psychology. Tooby and Cosmides (1989) have proposed a Swiss army knife metaphor in which the human mind comprises a varied collection of special-purpose modules evolved to solve specific and unrelated problems, the most numerous and crucial of which arose with early humans and their small-group social interactions. This focus on specific adaptive challenges and the evolved cognitive capacities for solving them is a necessary and important propaedeutic in the mostly evolution-free field of cognitive psychology. But, in practice, evolutionary psychologists have focused mainly on noncognitive (or only weakly cognitive) problems such as mate selection and incest avoidance. In terms of cognition, Tooby and Cosmides (2013) have been content simply to point out various ways in which human cognition shows the imprint of its evolutionary history in various domains. For example, in the domain of reasoning, humans solve some logical problems better if they are presented in social contexts similar to those from the environment of evolutionary adaptedness; in the domain of spatial cognition women have better spatial memories than men because they are adapted for plant gathering; and in the domain of visual attention humans pay special attention to the comings and goings of animals. So far, however, these theorists have given no comprehensive account of human cognitive modules in general, or of their unique aspects relative to other primates in particular.

There are several theories from this general perspective—that is, focusing on modularity and adaptedness—that make more systematic attempts to account for human cognitive uniqueness. First, Sperber (1996, 2000) argues that humans, like all animal species, possess a host of highly specific cognitive modules, such as snake detection and face recognition, as well as some more general modules, such as intuitive physics and intuitive psychology. These support what he calls intuitive beliefs (fast, resistant to evidence). What makes human cognition especially powerful is a kind of supermodule that enables individuals to entertain metarepresentations, that is to say, representations that not only cognitively represent the world but also represent others' or their own representations of the world. Individuals do this "propositionally" (compositionally and recursively), leading to what Sperber calls reflective beliefs (which may be formed either by having good reasons or by adopting the beliefs of others whom one trusts). If other animals engage in meta-representation at all, it is only in very rudimentary fashion, without compositionality and recursivity. This metarepresentational ability (Sperber actually thinks there may be three different metarepresentational modules) enables everything from cooperative (ostensive-inferential) communication, to teaching and cultural transmission, to reasoning by arguing with others. The capacity for metarepresentation coevolved with and interacts with a separate language module, which is also, obviously, uniquely human.

Carruthers (2006) gives an account of nonhuman primate cognition that involves representations and inferences, but he also stresses the limitations imposed by the "compartmentalization" of nonhuman primate cognitive modules. Human cognition is much more creative and flexible because, in the course of human evolution, additional modules were added, the most important of which are a mind-reading system (going beyond what apes do), a language-learning system, and a normative reasoning system. These modules can be applied simultaneously in the same situation, which creates some novelties, and in addition, humans evolved a disposition to imagine and rehearse action plans creatively in working memory, a capacity that enables all of the other modules to interact with one another more flexibly.

Mithen (1996) makes a systematic attempt to provide a modular theory of human cognitive evolution closely tied to the artifactual record. He makes a distinction between early humans and modern humans, noting that early humans were relatively limited cognitively, using the same tools everywhere over many millennia, with no symbolic behavior, and so forth. He explains this limitation by positing that early humans, like most animals, possessed several different cognitive modules that were not integrated with one another. Specifically, they had a technical intelligence with tools, a natural history intelligence with animals, and a social intelligence with conspecifics, none of which interacted any other module. With modern humans, we get symbolic capacities and language, which enabled the modules to work together, creating the kind of "cognitive fluidity" associated with modern human thinking.

What all of these more specific evolutionary psychology accounts have in common is the proposal that nonhuman primates, and perhaps even early humans, are dominated by highly compartmentalized modules, and this means that their cognitive processes are relatively narrow and inflexible. In contrast, human cognition is broader and more flexible because humans have modules, including some novel modules, that somehow work together or communicate with one another (via metarepresentation, symbols and language, or some horizontal processes such as creative imagination in working memory). The means that nonhuman animals (and perhaps early humans) operate only with system 1 intuitive inferences, whereas modern humans operate in addition with system 2 reflective inferences based on actual thinking. But this view—a kind of strict view of modularity for all animals except modern humans—is not compatible with the data on great ape thinking at all. There is no evidence that great apes operate only with highly compartmentalized modules, and indeed, chapter 2 presented evidence that they do not; they often use system 2 processes to think before they act in both the physical and social domains, in both cases using abstract representations, simple inferences, and protological paradigms (structured by physical causality or social intentionality). In our view, then, these attempts to both be true to modularity theory and simultaneously make room for human flexible thinking simply do not accord well with available empirical evidence.

It is also disconcerting to see how different are the specific modules that the different theorists posit—indeed, they often operate at very different levels of analysis (compare snake detection and face recognition with technical intelligence and normative reasoning). Perhaps a more systematic and comprehensive list could be compiled, but the real problem is that modularity theorists do not often ask the question of origins beyond seeking a single evolutionary function for a module (what it is "good for"). It is well known that in evolution new functions are often subserved by existing structures, perhaps put together in some new ways. Thus, for example, the proposed module for normative reasoning would almost certainly have been constructed out of earlier skills and motivations for such things as making individual inferences, conforming to others and the group, evaluating others and being sensitive to their evaluations, cooperative communication, and other skills. Looking at the architecture of contemporary human cognition for a single evolutionary function (via "reverse engineering") misses out on the dynamics of evolution, the way that new functions are created by cobbling together already existing structures as evolution proceeds. This dynamic means that there are deep relations among many cognitive functions via "common descent." A complex adaptive behavior such as collaborative foraging, for example, may comprise many component process such as fast running, accurate throwing, and skillful tracking—not to mention skills of joint intentionality—that may each have other adaptive functions, either on their own or in other complex behaviors. Once we get past narrowly defined problems with immediate and urgent fitness benefits (e.g., mate selection and predator detection), this hierarchical structure is crucial for understanding how different cognitive skills interrelate with others.

Our preference would thus be not to use the word *module*, which suggests a static architectural or engineering perspective. Rather, we would prefer the word *adaptation*, which suggests dynamic evolutionary processes. Adaptations may be quite narrowly targeted, and we ourselves have invoked the ethological notion of adaptive specialization (e.g., spiders building webs), which is very close in spirit to the notion of a module. But other adaptations may apply more broadly, either initially or through extensions over time. For example, great apes do not seem to be specifically adapted for tool use, as neither gorillas nor bonobos (and only some orangutan populations) use tools in the wild. But all great apes use tools, and quite adeptly, in appropriate circumstances in captivity. The adaptation thus seems to be more for causal understanding in the manipulation of objects, which can then be applied in the use of tools if the need arises for an individual (in contrast to some bird species, which seem to be specifically adapted for tool use).

Even more generally along these lines, the question arises whether there exist any truly domain-general horizontal abilities. (The metaphor is that specific content, like space or quantity, is vertical, whereas general processes like representation, memory, and inference are horizontal.) Some modularity theorists believe that seemingly horizontal abilities do not represent single domain-general processes; rather, each module has its own computational procedures that have nothing to do with those in other modules. Our view is that this again misses the importance of hierarchical organization in complex adaptations. Processes such as cognitive representation, inference, and self-monitoring may have evolved initially—in some ancient vertebrate ancestor—in the context of some fairly narrow behavioral specializations. But as new species have evolved, in the face of new and complex problems, these processes have been co-opted for use as subcomponents, as it were, in many different adaptations, some quite broad. This co-option process is especially important in highly flexible organisms such as great apes and humans, and indeed, wide-ranging occurrence of this process is a key component of cognitive flexibility.

Finally, we must also note that human skills and motivations for shared intentionality do not, in our view, represent typical cognitive adaptations occurring within individuals. Early humans had their own individual cognitive skills, but then they began attempting to coordinate with others toward joint goals with joint attention. Solving these coordination problems did not end the matter, however; rather, it opened up a whole new way of operating for early humans, especially the possibility of communicating referentially about basically everything in their experience with modified processes of representation and inference. The emergence of shared intentionality thus effected a restructuring, a transformation, a socialization, of all the processes involved in individual intentionality and thinking—an unusual, if not unprecedented, evolutionary event. This does not mean that humans do not also operate with many system 1 processes impervious to this transformation; they do, quite often making "gut decisions" about event probabilities, moral dilemmas, dangerous situations, and so forth (see, e.g., Gigerenzer and Selton, 2001; Haidt, 2012). But still, humans may consider and even communicate about all of these things in their system 2 thinking, even if this does not affect their eventual behavioral decision. Skills and motivations for shared intentionality thus transform the way that humans think about almost everything—because they can communicate about almost everything.

In any case, as noted at the outset, none of these various hypotheses—in any one of the three sets just reviewed—is a direct competitor to the shared intentionality hypothesis, as none of them focuses specifically on human thinking and its component processes. Each has captured its own segment of the truth about uniquely human cognition and thinking, we would argue, but the current account is both more comprehensive in covering all of the many different aspects of human thinking and more true to the way that evolutionary processes operate in cobbling together complex behavioral functions out of preexisting component processes. In addition, as we shall now see, the shared intentionality hypothesis also fits quite well with contemporary theories of the evolution of human sociality.

## Sociality and Thinking

There are a number of different theories of the evolution of human sociality, but they all agree on one thing: the general direction is one of ever more cooperation (at least until the rise of agriculture, cities, and stratified societies, some 10,000 years ago). As distinct from other great apes, early humans began mating via pair bonding, with the result that nuclear families became newly cooperating social units (Chapais, 2008). Relatedly, humans—again as opposed to other great apes—began various forms of cooperative childcare in which adults other than mothers cared for youngsters (Hrdy, 2009). This new form of childcare may have been a precursor to, but also may have occurred in conjunction with, collaborative foraging, as grandmothers and other females remained at home with the children while the healthiest females foraged and brought back the food to share—which turned the network of families involved into new cooperating units (Hawkes, 2003). And with the rise of modern humans, entire cultural groups—potentially encompassing whole clans or tribes with individuals who might not even know one another—became cooperating units as they competed with other human groups for valuable resources in cultural group selection (Richerson and Boyd, 2006).

How this trend toward cooperation might have interacted with humans' ever increasing cognitive competencies has been little explored, or even speculated about. There are two main exceptions. First, in support of the social brain hypothesis, Dunbar (1998) has documented across primate species a strong positive correlation between brain size (presumably reflecting cognitive complexity) and population size (presumably reflecting social complexity). Modern humans are the extreme case: human brain size and population size are both several times larger than those of their nearest great ape relatives. Gowlett et al. (2012) attempted to trace this relationship across human evolution and found an especially big jump in both brain size (as estimated by cranial volume) and estimated group size at around 400,000 years ago with *Homo heidelbergensis*—which is, of course, precisely at the time of our hypothesized first step in the evolution of human thinking via joint intentionality. However, group size is only a very gross indicator of social complexity (Dunbar focuses on the greater numbers of social relationships and reputations to be kept track of), and brain size is only a very gross indicator of cognitive capacities, so the social brain hypothesis gives us only a very general indication of the actual processes involved on either side of the correlation.

A more specific attempt to link human sociality and cognition is provided by Sterelny (2012), who has focused on human cooperation and its many facets, including cooperative childcare, cooperative foraging, and cooperative communication and teaching. The human cooperative lifestyle depends on individuals acquiring huge amounts of information during ontogeny—everything from how to track an antelope, to how to make a spear, to how the kinship relations of the group are constructed—and so the cooperative transmission of information from expert adults to novice children is crucial for individual survival. Humans have thus constructed learning environments within which their own offspring develop, which ensures that these offspring gain the information they need to perform critical subsistence activities such as toolmaking and collaborative foraging. Tomasello (1999) also offers a version of this theory, focusing especially on the ways in which human cognitive ontogeny is made possible by children acquiring the material and symbolic artifacts (including language) created by their forebears. In a generally similar vein, Levinson (2006) focuses on the uniquely human "interactive engine" of cooperative social engagement and how its evolution has created uniquely human forms of multimodal communication. Hrdy (2009) stresses that some of the adaptations involved here could have been for infant behavior itself, for example, special skills of cooperation and communication that enabled infants to navigate the newly complex world of multiple caregivers from an early age.

From the current perspective, these two accounts of the interrelation of human sociality and cognition are both useful and generally correct. But we have focused here specifically on the underlying processes of thinking involved. We have done this at a level of detail showing relatively precisely how specific problems in the coordination of action (collaboration) and the coordination of intentional states (cooperative communication) might have presented themselves to humans at two different evolutionary periods, and how humans might have solved them via new forms of thinking (employing new forms of cognitive representation, inference, and self-monitoring). Early humans needed not just to keep track of social relationships and transmit useful information to their young, but in addition and most immediately they also needed to meet the many and varied challenges of subsistence via social coordination—which they did by developing the many and varied skills and motivations of shared intentionality, including the ability to conceptualize situations for others recursively in cooperative and conventional communication. The indissociability of social coordination and human thinking is captured quite nicely by Sellars (1962/2007, p. 385), who writes: "Conceptual thinking is not by accident that which is *communicated* to others, any more than the decision to move a chess piece is by accident that which finds an expression in a move on a board between two people."

And so, by way of a general summary of our account, let us focus on the specific issue of the relation between sociality and thinking as it arises at each step of the proposed natural history. The main conclusions may be expressed in four very general propositions.

I. COMPETITION WITH GROUPMATES LED TO SOPHISTICATED FORMS OF NONHUMAN PRIMATE SOCIAL COGNITION AND THINKING WITHOUT HUMAN-LIKE FORMS OF SOCIALITY OR COMMUNICATION. Basic mammalian sociality is simply the motivation to live in a social group. Within-group competition engenders social relations of dominance and, along with other factors, affiliation. Great apes, and perhaps other primates, engage in more-than-average social competition and so have developed skills for understanding the goals and perceptions of others as a way of flexibly predicting their behavior. They also are especially skillful at manipulating physical causes in tool use and the intentional states of others in gestural communication. Great apes collaborate—that is, actually work together—very little, and when they do, it is best characterized as what Tuomela (2007) calls "group behavior in I-mode," as in chimpanzees' group hunting in which each individual is attempting to capture the monkey for itself. Great ape communication is almost exclusively about attempting to direct the recipient's attention and behavior in some desired ways, not to inform them of things useful to them. There are no human-like joint goals; there is no cooperative communication for coordinating actions.

Great ape cognition and thinking are adapted to this social, but not very cooperative, way of life. Great apes attend to situations relevant to their goals and values and, in certain problem situations, simulate or imagine the effects of various causes on the problem ahead of time before acting, as a way of making an effective behavioral decision. They do this with cognitive representations that are imagistic and schematic, understanding that "this is another one of those." They also understand in many cases how situations (and their components) relate to one another causally or intentionally, which enables them to simulate nonactual situations and make all kinds of causal and intentional inferences about them, including logical inferences organized into paradigms. For example, they infer not only "if X is present, then Y will be absent," but also "if there is only silence coming from here, then X must be there," and even "if X wants Y and perceives it in location Z, then she will go to location Z." These causal and intentional inferences also generate a kind of instrumental rationality in decision making, as the individual infers "if situation X obtains, then the best action to choose is Y." Great apes also self-monitor their own decision making not only by monitoring how outcomes match goals but also by monitoring the information available to them, and their confidence in it, before making a decision.

And so, the upshot is that great ape sociality has led to some remarkable skills of social cognition, to complement sophisticated skills of physical condition, what we have called skills of individual intentionality. But this form of sociality has not led to any transformations in the way that individuals conceptualize the world or think about problems in general. Individual intentionality has enabled great apes, and perhaps other nonhuman primates, to actually think about problems in specific situations, and to do this without any of humans' unique forms of sociality or communication. Individual intentionality and instrumental rationality may thus be considered general primate issue for "thought in a hostile world" (Sterelny, 2003.)

2. EARLY HUMAN COLLABORATIVE ACTIVITIES AND COOPERA-
TIVE COMMUNICATION—EMPLOYING NEW FORMS OF SOCIAL
COORDINATION—LED TO NEW FORMS OF HUMAN THINKING
WITHOUT EITHER CULTURE OR LANGUAGE. For more than 5 million
of the 6 million years that humans have been on their own evolutionary
pathway, their thinking was mainly ape-like (though their skills at making
tools may have enhanced their causal understanding). But then there was a
change in ecological conditions that forced some early humans to begin collaborating in new ways to obtain food. This made individuals interdependent
with one another in an especially urgent way. In mutualistic activities such as
these, communication could become fully cooperative since it was in the interest of each individual to coordinate with others toward their mutualistic
goal and to inform them of things useful to them in their role. And so were
born early humans who could survive and thrive only by collaborating and
communicating cooperatively with social partners.

Collaborative foraging created a number of difficult problems of social
coordination. The basic solution was to form with others joint goals to do
things together, to which both participants were jointly committed. This created the dual-level structure: joint goals with individual roles, along with
joint attention with individual perspectives. In the cooperative communication used to coordinate individuals' perspectives (and so actions) within these
activities—initially via pointing and pantomiming—the communicator was
committed to cooperation in the form of an honest informative act, and
communicator and recipient collaborated to ensure successful communication. The recipient followed the pointing gesture, or imagined the referent of
the pantomime, and then made an abductive inference from that to what,
given their common ground, the communicator intended to communicate.
The communicator, for his part, knew that this was what the recipient would
be doing and so attempted to conceptualize the situation for her in his choice
of referents—anticipating her perspective of his perspective on her perspective recursively—in a way that facilitated her abductive leap. Moreover, in
the special context of joint decision making, early human communicators
sometimes pointed out relevant situations to their partner that (implicitly)
provided reasons for them to jointly decide on a certain course of action based
on their common ground understanding of the causal and/or intentional
implications of the indicated situation.

To do all of this effectively required thinking of a type not possible for great apes and their individual intentionality: the communicator had to make judgments not only about his common conceptual ground with recipient but also about which aspects of the current situation the recipient would find both relevant and new—and so what kind of abductive inference she would make given different possible referential acts. Doing this led to what we have called second-personal thinking, comprising (1) cognitive representations that are perspectival and symbolic, (2) inferences that are recursively structured to include intentional states within intentional states, and (3) self-monitoring that incorporates the imagined social evaluation and/or comprehension of the collaborative and/or communicative partner. These changes all served to basically "cooperativize" great ape individual intentionality into a kind of second-personal joint intentionality and thinking.

And so, early humans' joint intentionality and second-personal thinking represented a radical break, a new type of relation between sociality and thinking. The cooperative and recursive sociality of early humans created an adaptive context requiring individuals, if they were to survive and thrive, to coordinate their actions and intentional states with others, which required them to "cooperativize" their cognitive representations, inferences, and self-monitoring, and so the processes of thinking that these enabled. Importantly for theories of the relation of sociality and thinking, this new type of second-personal thinking took place without conventionalization, culture, or language or anything else going beyond direct, second-personal, social engagements.

3. MODERN HUMAN PROCESSES OF CONVENTIONALIZED CULTURE AND LANGUAGE LED TO ALL OF THE UNIQUE COMPLEXITIES OF MODERN HUMAN THINKING AND REASONING. Modern humans faced some new social challenges due to increases in group sizes accompanied by competition among groups. For survival, modern human groups had to begin operating as relatively cohesive collaborative units, with various division-of-labor roles (see Wilson, 2012). This created the problem of how individuals could coordinate with in-group strangers, with whom they had no personal common ground. The solution was the conventionalization of cultural practices: everyone conformed to what everyone else was doing, and expected others to conform as well (and expected them to expect them to, etc.), which created a kind of cultural common ground that could be assumed of all members of the group (but not other groups). Modern humans' ways of com-

municating were conventionalized in this same way as well, which meant that individuals operated in a cultural common ground comprising a kind of group perspective and with conventionalized linguistic items and constructions that could be used effectively with anyone in the group.

This group-minded structuring of modern humans' activities and interactions, along with their conventional means of communication, meant that modern humans came to construct a kind of transpersonal, "objective" perspective on the world. Conventional communication became fully propositional, not only because of its conventional, normative, "objective" format and topic-focus structuring, but also because the speaker's communicative motives and epistemic/modal attitudes could be independently controlled in conventional signs, which meant that the propositional content was conceptualized independent of the motives and attitudes of particular individuals. Linguistic constructions enabled unprecedented creativity of conceptual combination, and moreover, they enabled full propositions representing a kind of generic, timeless, "objective" state of affairs, as in pedagogy ("It works like this") and the enforcement of social norms ("One must not do that"). Group-minded individuals thus constructed an "objective" world.

Conventional linguistic communication provided developing children with a preexisting representational system of alternative means of conceptualization, and everyone knew together in cultural common ground the available alternatives. This opened up a whole new world of both formal and pragmatic inferences. Processes of discourse aimed at effective communication encouraged communicators to make explicit many aspects of their own psychological processes left implicit in previous forms of communication (e.g., intentional states, logical operations), which enabled new ways of reflecting on thinking. In addition, cooperative argumentation for making joint decisions required that individuals make explicit their reasons and justifications to others in order to convince them of their truth; therefore, to be effective, they had to meet the group's normative expectations for rational discourse. Internalizing this reason-giving process meant that individuals now knew why, for what group-accepted reason, they were thinking what they were thinking. This process provided conceptual links between the individual's myriad thoughts and propositional representations, leading to a kind of holistic conceptual web. Each individual was also now practicing a kind of normative self-governance in which she, as emissary of the group to which she was a committed member, regulated her own actions and thoughts in terms of the group's normative standards.

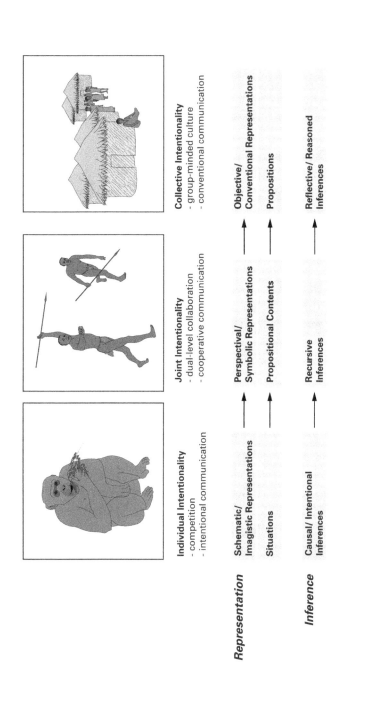

**Individual Intentionality**
- competition
- intentional communication

**Joint Intentionality**
- dual-level collaboration
- cooperative communication

**Collective Intentionality**
- group-minded culture
- conventional communication

*Representation*

Schematic/
Imagistic Representations → Perspectival/
Symbolic Representations → Objective/
Conventional Representations

Situations → Propositional Contents → Propositions

*Inference*

Causal/ Intentional
Inferences → Recursive
Inferences → Reflective / Reasoned
Inferences

*Self-Monitoring*

Cognitive
Self-Monitoring → Second-Personal
Self-Monitoring → Normative
Self-Governance

FIGURE 5.1  Summary of the shared intentionality hypothesis

And so, modern humans' creation of the various form of collective intentionality—comprising cultural conventions, norms, and institutions, including language—led to a kind of agent-neutral, "objective" thinking comprising conventional and objective representations; processes of inferring that were reasoned, reflective, and aimed at truth; and normative self-governance in which individuals monitored and adjusted their thinking to fit with that of the group. Culture and language, as agent-neutral conventional phenomena, thus provide another setting within which a new form of human sociality can lead to a new form of human thinking, specifically, objective-reflective-normative thinking.

From an evolutionary point of view, then, our overall argument is an extension of that of Maynard Smith and Szathmáry (1995): humans have created genuine evolutionary novelties via new forms of cooperation, supported and extended by new forms of communication. Further, this has led to new forms of cognitive representation, inference, and self-monitoring together constituting new forms of thinking. And humans have done this twice, the second step building on the first. Figure 5.1 summarizes the three component processes of human thinking at each of the three steps (i.e., including apes as step 0) of the shared intentionality hypothesis.

4. CUMULATIVE CULTURAL EVOLUTION LED TO A PLETHORA OF CULTURALLY SPECIFIC COGNITIVE SKILLS AND TYPES OF THINKING. All of these processes of joint and collective intentionality are universal in the human species. Most likely, the first step of joint intentionality evolved in Africa before the split between Neanderthals and modern humans and so characterized both species. The second step of collective intentionality likely evolved in a population of modern humans in Africa before they migrated out into other parts of the world after 100,000 years ago. But once they started migrating out and settling in highly variable local ecologies, differences in cultural practices became pronounced. Different human cultures created very different sets of particular cognitive skills, for example, for navigating across large distances, for building important tools and artifacts, and even for communicating linguistically. This meant that different cultures created, on top of their species-wide cognitive skills of individual, joint, and collective intentionality, many culturally specific cognitive skills and ways of thinking for their own local purposes.

Importantly, these culturally specific skills build on one another over historical time within a culture in a kind of ratchet effect, leading to cumulative cultural evolution. Because of humans' especially powerful skills of cultural learning, along with adult teaching and children's tendency to conform, the artifacts and practices of a culture acquire a "history." Individuals mediate their interactions with the world through the culture's artifacts and symbols from early in ontogeny (Vygotsky, 1978; Tomasello, 1999), thus absorbing something of the wisdom of the entire cultural group and its history. Cumulative cultural evolution is what enabled humans to conquer all kinds of otherwise uninhabitable places all over the globe.

As one dramatic example in the contemporary world, we may point to what are arguably the most abstract and complex forms of human thinking, that is, those involved in Western science and mathematics. The point here is that these forms of thinking are simply not possible without special forms of socially constructed conventions, namely, those in written form, that developed over historical time in Western culture. This point is stressed especially by Peirce (1931–1958) and is summarized in the classic text of modern logic by Lewis and Langford (1932, p. 4): "Had it not been for the adoption of the new and more versatile ideographic symbols, many branches of mathematics could never have been developed because no human mind could grasp the essence of their operations in terms of the phonograms of ordinary language." Many scholars of literacy would also argue that written language makes certain forms of reasoning, if not possible, at least more accessible (Olson, 1994). Writing also greatly facilitates metalinguistic thinking and the possibility to analyze, criticize, and evaluate our own linguistic communication, as well as that of others. Pictures and other graphic symbols used as communicative devices are collective representations that contribute to the process in important ways as well.

Those modern cultures that have created active communities of scientists, mathematicians, linguists, and other scholars are pretty much unthinkable without written language, written mathematical numerals and operations, and other forms of visually based and semipermanent symbols. Cultures that have not created and do not currently possess any of these kinds of graphic symbols cannot currently participate in these activities. This demonstrates quite clearly that many of the most complex and sophisticated human cognitive processes are indeed culturally and historically constructed. It also opens the possibility that some other human cognitive achievements are a kind of co-evolutionary mixture. Our own view would be that many of the complexities of human language are of this nature: built on universal cognitive processes but with culturally constructed concrete manifestations (Tomasello, 2008).

It is theoretically possible that this entire account applies not to human thinking in general but only to a kind of modularized thinking for collaboration and communication specifically (see Sperber, 1994, for something in this general direction). But this does not seem to be the case. Human perspectival and objective representations, recursive and reflective inferences, and normative self-monitoring—the constituents of uniquely human thinking—do not just go away when humans are not collaborating or communicating. On the contrary, they structure nearly everything that humans do, with the possible exception of sensory-motor activities. Thus, humans use recursive inferences in the grammatical structures of their languages, in mind-reading in non-communicative contexts, in mathematics, and in music, to name just the most obvious examples. Humans use perspectival and objective representations for thinking about everything, even in their solitary reveries, and they are engaged in normative self-monitoring whenever they are concerned about their reputation—which is pretty much all of the time. We might also recall here skills of relational thinking, which are products of dual-level collaboration but used more broadly, and skills of imagination and pretense, which are products of imagining in pantomime but are now used in all kinds of artistic creation. Collaboration and communication may play the crucial instigating roles in our story, but their effects on cognitive representation, inference, and self-monitoring extend much more broadly to basically all of humans' conceptually mediated activities.

Along these same lines, we should also be clear that the new forms of social cognition that this account proposes are not just modularized theory of mind skills. Rather, such things as perspectival representations, recursive inferences, and social self-monitoring evolved so that individuals could now understand the world in new ways by putting their heads together with others in acts of shared intentionality. Doing this requires more than just some specific cognitive skill aimed at some specific content domain, because coordinating actions and intentional states with others toward outside referents requires new ways of operating across the board. Skills and motivations for shared intentionality thus changed not just the way that humans think about others but also the way they conceptualize and think about the entire world, and their own place in it, in collaboration with others.

## The Role of Ontogeny

Although we have used ontogenetic data in various ways in this account, our focus here has not been on human ontogeny per se. It is thus important to make two key points about the role of ontogeny in the origins of uniquely human thinking.

First, although ontogeny does not have to recapitulate phylogeny, in the current case the relation between joint intentionality and collective intentionality is partly logical—one must have some skills for coordinating with other individuals before coordinating with the group—and so the ontogenetic ordering is basically the same as our hypothesized phylogenetic ordering (Tomasello and Hamann, 2012). In fact, however, things are more complicated than this because, as noted, young children are modern human beings, and they are exposed to many cultural artifacts, including a conventional language, practically from birth. But we would claim that, until around their third birthday, young children's social interactions with others are basically second-personal, not group based, and they do not fully understand how such things as language, artifacts, and social norms work as conventional creations.

And so, in the current view, the sequence is roughly this: Young children begin collaborating and communicating cooperatively with others with a second-personal orientation—through direct participation with specific other individuals—at around their first birthday. This includes engaging with others in joint attention, taking others' perspective in simple ways, and using the pointing gesture with others creatively (Carpenter et al., 1998; see Moll and Tomasello, in press, for a review). Importantly, this developmental timing is characteristic of children from a wide variety of cultural settings—including small-scale, nonliterate societies (Callaghan et al., 2011)—but it is not characteristic of chimpanzee ontogeny, even those raised by humans (Tomasello and Carpenter, 2005; Wobber et al., in press). This set of facts suggests a highly canalized and species-specific developmental pathway for the first emergence of skills of joint intentionality.

Skills of collective intentionality begin to emerge at around the third birthday. This is when young children first begin to understand social norms and other conventional phenomena as products of some kind of collective agreement. Thus, at around three years of age, young children do not just follow social norms but begin actively to enforce them on others (and to feel guilty when they break norms themselves). They do this in ways that demonstrate their understanding that particular norms apply only in particular contexts and only to individuals in the group who have conventionalized them. They also understand that some pieces of language, for example, common nouns, are conventional for everyone in the group, whereas proper names are conventional only for those who know the person (see Schmidt and Tomasello, 2012, for a review). Skills of collective intentionality have not been studied in any depth outside of Western, middle-class culture, and so the cross-cultural generality of this developmental timing is not known.

The second point about the role of ontogeny is this: neither joint nor collective intentionality would exist without it. This is true of many human traits, as the human species has evolved an extended ontogeny for all kinds of things that other species possess in mature, or almost mature, form at birth. Thus, whereas many small primates have brains that develop very rapidly in the first months of life, maturing in less than a year, and chimpanzees have brains that develop to maturity in only about five years, the human brain takes more than ten years to reach its fully mature adult volume (Coqueugniot et al., 2004). Because this extended ontogeny is highly risky both for youngsters and mothers, there must be offsetting advantages, presumably in terms of such things as especially flexible behavioral organization, cognition, and decision making—as well as time to master the local group's cultural artifacts, symbols, and practices (Bruner, 1972).

Human skills of joint and collective intentionality thus come into existence during an extended ontogeny in which the child and her developing brain are in constant interaction with the environment, especially the social environment. Our hypothesis is that they would not come into existence without this interaction. To make the point as concretely as possible, let us invoke a thought experiment we have used before, and then add a novel twist. Imagine a child born on a desert island, miraculously kept alive and healthy until adulthood, but all alone. The hypothesis is that this child, as an adult, would not have skills of either joint or collective intentionality. This social isolate could not, as an adult, enter a human group and start collaborating by forming joint goals with individual roles, or communicating cooperatively in the context of joint attention with individual perspectives. This individual would thus not develop during its isolated lifetime second-personal thinking with perspectival and symbolic representations, recursive inferences, and social self-monitoring. How could he develop an appreciation for perspectives that differ from his own without in fact experiencing different perspectives? How could he develop socially recursive inferences without any social partners with whom to communicate? How could he worry about others evaluating him if no such others existed? No, skills of shared intentionality are not simply innate, or maturational; they are biological adaptations that come into existence as they are used during ontogeny to collaborate and communicate with others.

This thought experiment might be called Robinson Crusoe, as the child is alone on the desert island. But now imagine a Lord of the Flies scenario. In this case it would be multiple infants born and growing to maturity on a desert island, with no one to interact with but each other. Perhaps surprisingly, the hypothesis in this case is that these children would indeed have the kind of social interactions necessary for developing joint intentionality—but not collective intentionality. That is to say, these orphaned peers should develop, through their social interactions with one another, skills of second-personal, recursive sociality. They would find ways to collaborate with one another with joint goals and attention, communicate with one another taking different perspectives (by pointing and pantomiming), and monitor their behavior through the eyes of their interdependent partner. To develop in this way, sophisticated adults and all of their cultural paraphernalia are not necessary.

But we do not think that peer interaction alone would be sufficient for our orphaned peers to develop skills of collective intentionality in their lifetimes. They might develop on their own some conventions and norms of some kind, as skills of joint intentionality plus imitation might be sufficient, and they might create something like a culture over many generations. But they would not develop during their own lifetimes a full-fledged culture or conventional language, as these take multiple generations over historical time to develop. And the same could be said of cultural institutions with standing status functions, like chiefs and money. In general, fully fledged skills of collective intentionality and agent-neutral thinking require, on our hypothesis, ontogeny in the midst of a preexisting cultural collective with preexisting conventions, norms, and institutions, including a conventional language. How could one become group minded, represent things objectively, and regulate one's behavior and reasoning by the cooperative and communicative norms of the social group if there was in fact no social group that antedated one's social and cognitive development? No, skills of collective intentionality are not simply innate or maturational either; they are biological adaptations that come into existence only through an extended ontogeny in a collectively created and transmitted cultural environment—which takes multiple generations to emerge. In this case, then, adults and all of their cultural paraphernalia are indeed necessary for the ontogenetic development of skills of collective intentionality.

It is not incoherent to believe that all of the cognitive and thinking skills we have described could be built-in—to the degree that our wild child or orphaned peers could, if discovered as adults, immediately display perfectly mature forms of uniquely human thinking at both levels. It is just, in our view, highly unlikely. Humans biologically inherit their basic capacities for constructing uniquely human cognitive representations, forms of inference, and self-monitoring, from out of their collaborative and communicative interactions with other social beings. Absent a social environment, these capacities would wither away from disuse, like the capacity for vision in a person born and raised completely in darkness.

One could, in principle, collect data on the role of ontogeny in the emergence of uniquely human thinking—but only if one had no moral scruples. One would have to be prepared to randomly assigned newborn children to different rearing environments. Natural experiments, such as the feral child Victor of Aveyron and other "wolf" children, are not definitive on the question for many reasons, not the least of which is that some of these children could have been abandoned by their parents precisely because they were not normally functioning (Candland, 1995)—and none of them was tested for the relevant cognitive skills either. Some interesting indirect evidence for the important role of a human-like social environment is provided by so-called enculturated apes. When apes are raised by humans in the midst of all kinds of human-like social interaction and artifacts, they do not develop more human-like skills of physical cognition (e.g., space, object permanence, tool use), but they do develop more human-like skills of imitation and communication (Call and Tomasello, 1996; Tomasello and Call, 2004). The significance of these findings for human ontogeny are, however, not straightforward.

In any case, while everyone will continue to be fascinated by the question of wild children and how much and what kinds of social experience are necessary for humans to develop their unique forms of cognition and thinking, the question is very likely to remain a deep mystery for the foreseeable future. In the meantime, our hypothesis is that, like many human adaptations, adaptations for shared intentionality are built to grow and flourish only in the midst of rich social and cultural nourishment of particular kinds.

# 6

## Conclusion

In both the evolution of thought in the history of mankind,
and the evolution of thought in an individual, there is a stage at
which there is no thought followed by a subsequent stage at which
there is thought. . . . What we lack is a satisfactory vocabulary
for describing the intermediate steps.

—DONALD DAVIDSON, *SUBJECTIVE, INTERSUBJECTIVE, OBJECTIVE*

At least since Aristotle, human beings have wondered how they differ from other animal species. But for almost all of that time the appropriate information for making this comparison was not available—most important, because for the first several thousand years of Western civilization there were no nonhuman primates in Europe. Aristotle and Descartes could readily posit things like "only humans have reason" or "only humans have free will" because they were comparing humans to birds, rats, various domesticated animals, and the occasional fox or wolf.

In the nineteenth century nonhuman primates, including great apes, came to Europe via newly created zoological gardens. Darwin himself was dumbstruck by his encounter in 1838 with an orangutan named Jenny at the London Zoo (whom Queen Victoria termed "disagreeably human"). After publication of the *Origin of Species* some twenty-one years later, and the *Descent of Man* some twelve years after that, the differences between humans and other animals—as now represented by our closest living relatives—became much more difficult to pinpoint. Many philosophers reacted by simply defining away the problem: thinking is a process that takes place in, and only in, the medium of language, and so other animal species cannot think by definition (the most prominent modern proponents being Davidson [2001] and Brandom [1994]). Recent research on great ape cognition and thinking, such as that reviewed here, should already be undermining this "radical discontinuity" view. Great apes cognitively represent the world in abstract format, they make complex causal and intentional inferences with logical structure, and they seem to know, at least in some sense, what they are doing while they are doing it. Although this may not be fully human thinking, for sure it has some key components.

But the problem is deeper than finding a line of demarcation. The point is that the great ape species alive today are arbitrarily far from humans; it is just a matter of who survived and who did not. So what if we discovered, in some remote jungle, surviving members of the species *Homo heidelbergensis* or *Homo neanderthalensis*? How would we decide whether they possess fully human thinking—yes or no—given that, in all probability, they would be somewhere in between contemporary humans and great apes? Even more radical, what if we discovered some earlier side branches from the human evolutionary tree who had their own ways of doing and thinking about things, overlapping only partially with modern human thinking? Perhaps these creatures never developed pointing and so did not evolve skills of recursive inferring. Or perhaps they never imitated at a level sufficient for pantomime and so did not symbolize their experience for others gesturally. Or perhaps they collaborated but did not care about others' evaluations and so did not become socially normative. Or perhaps they never had situations in which they had to make group decisions and so never came to offer one another reasons and justifications for their assertions. Our question is what would these creatures' version of thinking look like if it skipped a key ingredient (along with all of its cascading effects) of the modern human version? We might end up with something sharing many features with modern human thinking but having its own unique features as well. The point is that, considered evolutionarily, human thinking is not a monolith but a motley—and it could have turned out other than it did.

What we have done in the current natural history is to imagine one possible "missing link" in the evolution of human thinking from great apes to modern humans based on selected aspects of the way of life of contemporary hunter-gatherers and selected aspects of the thinking of young children (accompanied by a few, admittedly indeterminant, paleoanthropological facts). But, importantly, our claim is not just that such an intermediate step can be imagined and that it probably occurred, but that it was *necessary*. It was necessary because one cannot even imagine going directly from ape-like competitive interactions and imperative communication to modern human culture and language with no evolutionary intermediary. Human culture and language are simply conventionalizations of existing social interactions, and to provide the appropriate raw material, these interactions had to have been already highly cooperative. Put in terms of our two historical strands, we cannot get to processes of culture and language (as invoked by Vygotsky and other culture theorists) without some kind of already existing and already cooperative social infrastructure (as described by Mead, Wittgenstein, and other social infrastructure theorists). We thus need our middle step—we would be very happy with multiple middle steps, if ours could be broken down further—to prepare the way for culture and language and all of their uniquely powerful structuring of human thinking. This intermediate step does not solve Davidson's (1982) problem of a common theoretical vocabulary that spans from "no thought" to "thought," but it does narrow the distance to be traversed by any one step significantly.

In any case, no matter the precise number of steps, our account presupposes that to understand uniquely human thinking we must situate it in its evolutionary context. Wittgenstein (1955, no. 132) says about language that "the confusions which occupy us arise when language is like an engine idling, not when it is doing its work." It seems to us that many of the perplexities of human thinking pointed out by philosophers arise precisely when we attempt to understand it in the abstract, outside of its functioning in solving adaptive problems. It is natural to do this in the contemporary world because so much contemporary thinking is, in some sense, idling. But uniquely human thinking was almost certainly selected evolutionarily for its role in organizing and regulating adaptive actions, and so to understand it fully we must identify the relevant problems. If creatures from outer space came across a complex human artifact such as a traffic light, idling, they could dissect it and analyze its structure forever and not understand why it behaves in the way that it does. The wires and lights by themselves could never reveal (not even with the help of an fMRI) why the red light on one side activates only when the green light on the other side activates. To understand these actions, we must first understand traffic and how the traffic light was designed to solve the specific problems created by traffic. In the case of biological structures—and this is a central lesson of evolutionary psychology, of course—there is the additional possibility that they evolved to serve one set of functions at an earlier time, and now they serve a different set. In any case, the current proposal is that to understand the way that contemporary humans think, we must understand how human thinking evolved to meet the specific evolutionary challenges that early and modern humans faced as they moved toward ever more cooperative ways of making a living.

It is certain that some parts of our evolutionary story are incomplete. The main problem is that collaboration, communication, and thinking do not fossilize, and so we will always be in a position of speculation about such behavioral phenomena, as well as the specific events that were critical to their evolution. Most crucial, we do not know how much contemporary great apes have changed from their common ancestor with humans because there are basically no relevant fossils from this era. Furthermore, our intermediary step of early humans very likely had much more of a gradual evolution than described here; indeed, it is not even clear that *Homo heidelbergensis* was a separate species at all. And we have given only cursory attention to humans after agriculture and all of the complexities arising from the intermixing of cultural groups, from literacy and numeracy, and from institutions such as science and government. And so our attempt is less of an explicitly historical exercise than an attempt to carve nature at some of its most important joints, specifically, at some of its most important evolutionary joints.

A list of open questions at this point would be quite long. But two particularly big ones are these: First is the nature of the jointness or collectivity or "we-ness" that characterizes all forms of shared intentionality. Many theorists subscribe to something like an irreducibility thesis (e.g., Gallotti, 2012) in which such things as joint attention and shared conventions are irreducibly social phenomena, and attempting to capture them in terms of the individuals involved, and what is going on in their individual heads, is doomed to failure. Our view is that shared intentionality is indeed an irreducibly social phenomena in the moment—joint attention only exists when two or more individuals are interacting, for example—but at the same time we may ask the evolutionary or developmental question of what does the individual bring to the interaction that enables her to engage in joint attention in a way that other apes and younger children cannot. And so for us this means that something like recursive mind-reading or inferring—still not adequately characterized, and in most instances fully implicit—has to be a part of the story of shared intentionality. From the individual's point of view, shared intentionality is simply experienced as a sharing, but its underlying structure, reflecting its evolution, is that each participant in an interaction can potentially take the perspective of others taking her perspective taking their perspective, and so forth for at least a few levels. But this, as they say, is a point on which reasonable people may disagree.

A second open question is how and why modern humans reify and objectify what are essentially socially created entities. Money is not just a piece of paper but legal tender, and Barack Obama is not just a person living in a large white house but commander in chief—because we act and talk as if they are these things. We also reify such things as morality, arguing not about the moral norms of different social groups, including those shared by all human groups, but rather about what is the "right" and "wrong" way to do things, where right and wrong are considered objective features of the world. And nowhere is this tendency stronger than in language, where everyone has a tendency—correctable but only with much effort—to reify the conceptualizations codified in our own natural language. About all of these things, we are like the young child who says that even if long ago everyone agreed to call the striped feline in front of us a "gazzer," it would not be right to do so because, well, "It's a tiger." Our own view is that such objectifying tendencies could come only from the kind of agent-neutral, group-minded perspective that imagines things from the view of any one of us, the view of any rational person, the view from nowhere, in the context of a world of social and institutional realities that antedate our own existence and that speak with an authority larger than us. This is the authoritative voice that lies behind the use of genericized linguistic expressions in norm enforcement ("That is wrong") and pedagogy ("It works like this"), and it determines, in large part, what we consider real. But, again, this is a point on which reasonable people may disagree.

Despite these gaping questions, and others, we cannot conceive any comprehensive theory of the origins of uniquely human thinking that is not fundamentally social in character. To be as clear as possible: we are not claiming that all aspects of human thinking are socially constituted, only the species-unique aspects. It is an empirical fact that the social interaction and organization of great apes and humans are hugely different, with humans being much more cooperative in every way. We find it difficult in the extreme to believe that this is unrelated to the huge differences in cognition and thinking that also separate great apes from humans, especially when we focus on the details. What nonsocial theory can explain such things as cultural institutions, perspectival and conventional conceptualizations in natural languages, recursive and rational reasoning, objective perspectives, social norms and normative self-governance, and on and on? These are all coordinative phenomena through and through, and it is almost inconceivable that they arose evolutionarily from some nonsocial source. Something like the shared intentionality hypothesis just must be true.

# Notes

## 2. Individual Intentionality

1. Importantly, complex organisms embody hierarchies of control systems, so that most of their actions are attempts to regulate multiple goals simultaneously at multiple levels (e.g., the same act is simultaneously attempting to place left foot in front of right, pursue a prey, feed the family, etc.).

2. This account is related to the notion of Gibsonian affordances, but it is much broader in including not only direct opportunities for the self's concrete actions but also situations that are relevant to the organism in many more indirect ways. In addition, we should also acknowledge that all organisms are hardwired to attend to some things as naturally salient (e.g., for humans, loud noises) because of potential relevance to biological "goals" and "values" (so-called "bottom-up" processes of attention).

3. In none of these studies did chimpanzees understand and predict that, when a dominant was not just ignorant but had a false belief, she would reliably go to the place where she (falsely) thought the food was located. They treated ignorance and false belief as the same (see also Kaminski et al., 2008; Krachun et al., 2009, 2010; chapter 3 discusses this distinction further).

## 3. Joint Intentionality

1. Contemporary human foragers are not good models for the early humans we are imagining here, as they have gone through both steps of our evolutionary story

and so live in cultures with social norms, institutions, and languages. Moreover, contemporary foragers have tools and weapons that make individual foraging (then sharing at the end) feasible, whereas the early humans we are imagining here had more primitive weapons and so needed to work together.

2. Of course, contemporary human societies are also full of selfishness and non-cooperation, not to mention cruelty and war. Much of this is generated by conflicts between people from different groups (however this is defined) and concerns competition for private property and the accumulation of wealth that began only in the last 10,000 years or so, after the advent of agriculture, that is, after humans had spent many millennia as small-group collaborative foragers.

3. Davidson is actually concerned with a special kind of perspective, namely, belief: a cognitive representation of the world that the subject knows might be in error. His claim is that a necessary condition for the notion of error is a social situation in which I and another person simultaneously focus on the same object or event simultaneously yet differently, what we have called perspective. But the notion of error introduces an additional consideration because it privileges one of the perspectives as accurate (and the other as in error), and this requires some notion of an "objective" perspective. This notion of objectivity—and so the notion of belief—will not be available to humans until the next step in our story when agent neutral perspectives are possible (see chapter 4).

4. The possibility of lying meant that recipients had to practice "epistemic vigilance" (Sperber et al., 2010). And so the notion of true propositions also arose from the comprehension side of the interaction, as comprehenders attempted to distinguish truthful from deceptive communicative acts.

5. For purposes of simplicity, the terminology here—referential acts underlain by communicative intentions—is slightly different from that of Tomasello (2008). What is here called the communicative intention comprises what was there called the social intention in the context of the Gricean communicative intention.

6. Some researchers think that this characterization of children's early communication via pointing is too cognitively rich (see, e.g., Gomez, 2007; Southgate et al., 2007) and that infants are actually doing something simpler.

7. Said another way, it is one thing to throw an object *at* another person (which, coincidently, many apes do), but it is quite another to throw something *to* someone in anticipation of her task of catching (Darwall, 2006)—which is what the communicator does, metaphorically, in human cooperative communication.

8. Some researchers have claimed that some great ape intention-movements are actually functioning iconically, for example, when one gorilla ritualistically pushes another in a direction in a sexual or play context (Tanner and Byrne, 1996). But these are most likely garden-variety ritualized behaviors that appear to humans to be iconic because they derive from attempts to actually move the body of the other in the desired direction—they are not functioning iconically for the apes themselves.

9. Some contemporary cultures have more than one (e.g., pointing with the index finger and pinky extended simultaneously for a certain subclass of situations), but the presumption is that those are derived from the original primordial index finger pointing, with which all children begin.

10. We have until this point discussed only "propositional contents" in the sense of fact-like situations that are expressed in cooperative communication. By the term *proposition* we mean a communicative act expressed as a fully articulated act of conventional linguistic communication.

## 4. Collective Intentionality

1. Children also find it difficult for some time to comprehend situations in which objective reality is unaffected by the fact that we humans may describe it from different, even conflicting, perspectives, for example, situations in which there is an undisturbed objective reality despite the fact that this entity is simultaneously a dog an animal and a pet (see Moll and Tomasello, in press).

2. This is not unlike the way that some motivated linguistic forms, such as metaphors, become opaque ("dead") across historical time as new learners are ignorant of the original motivation.

3. A number of unusual situations in the contemporary world have illustrated the process, at least in broad outline. Most spectacular is the case of Nicaraguan Sign Language. A number of young deaf individuals each had their own kind of pidgin signing, or home sign, with very little grammatical structuring, that they used with their hearing families. But soon after they were brought together into a community—within three "generations"—their various idiosyncratic home signs turned into a system of conventionalized signs used in numerous constructions with all kinds of grammatical organization (Senghas et al., 2004). A very similar process was observed in the birth of Al-Sayyid Bedouin Sign Language (Sandler et al., 2005), and indeed, somewhat similar processes have been at least indirectly observed in many cases in which spoken pidgin languages have turned into creoles and full languages (Lefebvre, 2006). What seems to happen is that pidgin communication (or home sign) works well among family members, coworkers, and others with very strong common ground, typically in highly restricted and recurrent situations such as mealtime or a work task. But especially as a wider community of communicators and communicative situations must be accommodated, this process breaks down, and new grammatical means must be found to help recipients to reconstruct the events and participants (and their roles) in the intended referential situation. Communicator and recipient then work together further until there is comprehension, and successful grammatical solutions are repeated and imitated and so conventionalized in the community.

4. Participants and events in situations may be linguistically indicated at many different levels of specificity, depending on the common ground between communicator

and recipient (Gundel et al., 1993). Pronouns are used to indicate entities already well established in common ground, whereas nouns with relative clauses are used for new entities that the recipient may identify using our common ground (e.g., "the man we saw yesterday"). In addition, many languages have determiners such as *the* and *a*, which specifically indicate whether something is or is not in our common ground in the current communicative interaction. Events are typically grounded in the current communicative interaction by specifying when they occurred or will occur relative to, ultimately, now (i.e., via tense). This way of specifying referents thus leads to the kind of hierarchical tree structures diagrammed in traditional linguistic analyses, as the different linguistic items of a noun phrase or a verbal complex, each with its own function, are used together, collaboratively as it were, toward the overall goal of indicating a particular participant or event in the referential situation.

5. Sandler et al. (2005) provide a very interesting description of how successive generations of the newly created Al-Sayyid Bedouin Sign Language conventionalized speaker motive and attitude, mostly by conventionalizing slightly exaggerated facial expressions. Thus, across generations signers came to use conventionalized facial expressions to signal such things as "the illocutionary force of an utterance, such as assertions vs. questions" (p. 31)—as in mature sign languages. In addition, communicators in later but not earlier generations came to symbolize conventionally their various modal and epistemic attitudes, such things as necessity, possibility, uncertainty, or surprise.

# References

Alvard, M. 2012. Human sociality. In J. Mitani, ed., *The evolution of primate societies.* (pp. 585–604). Chicago: University of Chicago Press.

Bakhtin, M. M. 1981. The dialogic imagination (trans. C. Emerson and M. Holquist). In M. Holquist, ed., *Four essays.* Austin: University of Texas Press.

Barsalou, L. W. 1983. Ad hoc categories. *Memory and Cognition, 11,* 211–227.

———. 1999. Perceptual symbol systems. *Behavioral and Brain Sciences, 22,* 577–609.

———. 2005. Continuity of the conceptual system across species. *Trends in Cognitive Sciences, 9,* 309–311.

———. 2008. Grounded cognition. *Annual Review of Psychology, 59,* 617–645.

Behne, T., M. Carpenter, and M. Tomasello. 2005. One-year-olds comprehend the communicative intentions behind gestures in a hiding game. *Developmental Science, 8,* 492–499.

Behne, T., U. Liszkowski, M. Carpenter, and M. Tomasello. 2012. Twelve-month-olds' comprehension and production of pointing. *British Journal of Developmental Psychology, 30*(3), 359–375.

Bennett, M., and F. Sani. 2008. Children's subjective identification with social groups: A self-stereotyping approach. *Developmental Science, 11,* 69–78.

Bermudez, J. 2003. *Thinking without words.* New York: Oxford University Press.

Bickerton, D. 2009. *Adam's tongue.* New York: Hill and Wang.

Boehm, C. 2012. *Moral origins.* New York: Basic Books.

Boesch, C. 2005. Joint cooperative hunting among wild chimpanzees: Taking natural observations seriously. *Behavioral and Brain Sciences, 28,* 692–693.

Boesch, C., and H. Boesch 1989. Hunting behavior of wild chimpanzees in the Taï National Park. *American Journal of Physical Anthropology, 78,* 547–573.

Brandom, R. 1994. *Making it explicit: Reasoning, representing, and discursive commitment.* Cambridge, MA: Harvard University Press.

———. 2009. *Reason in philosophy: Animating ideas.* Cambridge, MA: Harvard University Press.

Bratman, M. 1992. Shared cooperative activity. *Philosophical Review, 101*(2), 327–341.

Brownell, C. A., and M. S. Carriger. 1990. Changes in cooperation and self-other differentiation during the second year. *Child Development, 61,* 1164–1174.

Bruner, J. 1972. The nature and uses of immaturity. *American Psychologist, 27,* 687–708.

Bullinger, A., A. Melis, and M. Tomasello. 2011a. Chimpanzees prefer individual over cooperative strategies toward goals. *Animal Behaviour, 82.* 1135–1141.

———. 2013. Bonobos, *Pan paniscus,* chimpanzees, *Pan troglodytes,* and marmosets, *Callithrix jacchus,* prefer to feed alone. *Animal Behavior, 85,* 51–60.

Bullinger, A., E. Wyman, A. Melis, and M. Tomasello. 2011b. Chimpanzees coordinate in a stag hunt game. *International Journal of Primatology, 32,* 1296–1310.

Bullinger, A., F. Zimmerman, J. Kaminski and M. Tomasello. 2011c. Different social motives in the gestural communication of chimpanzees and human children. *Developmental Science, 14,* 58–68.

Buttelmann, D., M. Carpenter, J. Call, and M. Tomasello. 2007. Enculturated apes imitate rationally. *Developmental Science, 10,* F31–38.

Buttelmann, D., M. Carpenter, and M. Tomasello. 2009. Eighteen-month-old infants show false belief understanding in an active helping paradigm. *Cognition, 112*(2), 337–342.

Call, J. 2001. Object permanence in orangutans (*Pongo pygmaeus*), chimpanzees (*Pan troglodytes*), and children (*Homo sapiens*). *Journal of Comparative Psychology, 115,* 159–171.

———. 2004. Inferences about the location of food in the great apes. *Journal of Comparative Psychology, 118*(2), 232–241.

———. 2006. Descartes' two errors: Reasoning and reflection from a comparative perspective. In S. Hurley and M. Nudds, eds., *Rational animals.* (pp. 219–34). Oxford: Oxford University Press.

———. 2010. Do apes know that they can be wrong? *Animal Cognition, 13,* 689–700.

Call, J., and M. Tomasello. 1996. The effect of humans on the cognitive development of apes. In A. E. Russon, K. A. Bard, and S. T. Parker, eds., *Reaching into
thought* (pp. 371–403). New York: Cambridge University Press.

———. 2005. What chimpanzees know about seeing, revisited: An explanation of
the third kind. In N. Eilan, C. Hoerl, T. McCormack, and J. Roessler, eds.,
*Joint attention: Communication and other minds* (pp. 45–64). Oxford: Oxford
University Press.

———. 2007. *The gestural communication of apes and monkeys.* Mahwah, NJ:
Lawrence Erlbaum.

———. 2008. Does the chimpanzee have a theory of mind: 30 years later. *Trends in
Cognitive Science, 12,* 87–92.

Callaghan, T., H. Moll, H. Rakozcy, T. Behne, U. Liszkowski, and M. Tomasello.
2011. *Early social cognition in three cultural contexts.* Monographs of the Society
for Research in Child Development 76(2). Boston: Wiley-Blackwell.

Candland, D. K. 1995. *Feral children and clever animals: Reflections on human
nature.* Oxford: Oxford University Press.

Carey, S. 2009. *The origin of concepts.* New York: Oxford University Press.

Carpenter, M., K. Nagel, and M. Tomasello 1998. *Social cognition, joint attention, and
communicative competence from 9 to 15 months of age.* Monographs of the Society for
Research in Child Development 63(4). Chicago: University of Chicago Press.

Carpenter, M., M. Tomasello, and T. Striano. 2005. Role reversal imitation in 12
and 18 month olds and children with autism. *Infancy, 8,* 253–278.

Carruthers, P. 2006. *The architecture of the mind.* Oxford: Oxford University Press.

Carruthers, P., and M. Ritchie. 2012. The emergence of metacognition: Affect and
uncertainty in animals. In M. Beran et al., eds., *Foundations of metacognition.*
(pp. 211–37). New York: Oxford University Press.

Chapais, B. 2008. *Primeval kinship: How pair-bonding gave birth to human society.*
Cambridge, MA: Harvard University Press.

Chase, P. 2006. *The emergence of culture.* New York: Springer.

Chwe, M. S.-Y. 2003. *Rational ritual: Culture, coordination and common knowledge.*
Princeton, NJ: Princeton University Press.

Clark, H. 1996. *Uses of language.* Cambridge: Cambridge University Press.

Collingwood, R. 1946. *The idea of history.* Oxford: Clarendon Press.

Coqueugniot, H., J.-J. Hublin, F. Veillon, F. Houet, and T. Jacob. 2004. Early
brain growth in Homo erectus and implications for cognitive ability. *Nature,
231,* 299–302.

Corbalis, M. 2011. *The recursive mind.* Princeton, NJ: Princeton University Press.

Crane, T. 2003. *The mechanical mind: A philosophical introduction to minds,
machines and mental representation.* 2nd ed. New York: Routledge.

Crockford, C., R. M. Wittig, R. Mundry, and K. Zuberbuehler. 2011. Wild chimpanzees inform ignorant group members of danger. *Current Biology, 22,* 142–146.

Croft, W. 2001. *Radical construction grammar.* Oxford: Oxford University Press.

Csibra, G., and G. Gergely. 2009. Natural pedagogy. *Trends in Cognitive Sciences, 13,* 148–153.

Custance, D. M., A. Whiten, and K. A. Bard. 1995. Can young chimpanzees imitate arbitrary actions? Hayes and Hayes (1952) revisited. *Behaviour, 132,* 839–858.

Darwall, S. 2006. *The second-person standpoint: Respect, morality, and accountability.* Cambridge, MA: Harvard University Press.

Darwin, C. 1859. *The origin of species.* London: John Murray.

———. 1871. *The descent of man.* London: John Murray.

Davidson, D. 1982. Rational Animals. *Dialectica, 36,* 317–327.

———. 2001. *Subjective, intersubjective, objective.* Oxford: Clarendon Press.

Dean, L. G., R. L. Kendal, S. J. Schapiro, B. Thierry, and K. N. Laland. 2012. Identification of the social and cognitive processes underlying human cumulative culture. *Science, 335,* 1114–1118.

Dennett, D. 1995. *Darwin's dangerous ideas.* New York: Simon and Schuster.

de Waal, F. B. M. 1999. Anthropomorphism and anthropodenial: Consistency in our thinking about humans and other animals. *Philosophical Topics, 27,* 255–280.

Diesendruck, G., N. Carmel, and L. Markson. 2010. Children's sensitivity to the conventionality of sources. *Child Development, 81,* 652–668.

Diessel, H., and M. Tomasello. 2001. The acquisition of finite complement clauses in English: A usage based approach to the development of grammatical constructions. *Cognitive Linguistics, 12,* 97–141.

Donald, M. 1991. *Origins of the modern mind.* Cambridge, MA: Harvard University Press.

Dunbar, R. 1998. The social brain hypothesis. *Evolutionary Anthropology, 6,* 178–190.

Engelmann, J., E. Herrmann, and M. Tomasello. 2012. Five-year olds, but not chimpanzees, attempt to manage their reputations. *PLoS ONE, 7*(10), e48433.

Engelmann, J., H. Over, E. Herrmann, and M. Tomasello. In press. Young children care more about their reputations with ingroup than with outgroup members. *Developmental Science.*

Evans, G. 1982. The varieties of reference. In J. McDowell, ed., *The varieties of reference.* (pp. 73–100). Oxford: Oxford University Press.

Fletcher, G., F. Warneken, and M. Tomasello. 2012. Differences in cognitive processes underlying the collaborative activities of children and chimpanzees. *Cognitive Development, 27,* 136–153.

Fragaszy, D., P. Izar, and E. Visalberghi. 2004. Wild capuchin monkeys use anvils and stone pounding tools. *American Journal of Primatology, 64,* 359–366.

Gallotti, M. 2012. A naturalistic argument for the irreducibility of collective intentionality. *Philosophy of the Social Sciences, 42*(1), 3–30.

Geertz, C. 1973. *The interpretation of cultures.* New York: Basic Books.

Gentner, D. 2003. Why we're so smart. In D. Gentner and S. Goldin-Meadow, eds., *Language in mind: Advances in the study of language and thought* (pp. 195–235). Cambridge, MA: The MIT Press.

Gergely, G., H. Bekkering, and I. Király. 2002. Rational imitation in preverbal infants, *Nature, 415,* 755.

Gigerenzer, G., and R. Selton. 2001. *Bounded rationality: The adaptive toolbox.* Cambridge, MA: The MIT Press.

Gilbert, M. 1983. Notes on the concept of social convention. *New Literary History, 14,* 225–251.

———. 1989. *On social facts.* London: Routledge.

———. 1990. Walking together: A paradigmatic social phenomenon. *Midwest Studies in Philosophy, 15,* 1–14.

Gilby, I. C. 2006. Meat sharing among the Gombe chimpanzees: Harassment and reciprocal exchange. *Animal Behaviour, 71*(4), 953–963.

Givón, T. 1995. *Functionalism and grammar.* Amsterdam: J. Benjamins.

Goeckeritz, S., M. Schmidt, and M. Tomasello. Unpublished manuscript. How children make up and enforce their own rules.

Goldberg, A. 1995. *Constructions: A construction grammar approach to argument structure.* Chicago: University of Chicago Press.

———. 2006. *Constructions at work.* Oxford: Oxford University Press.

Goldin-Meadow, S. 2003. *The resilience of language: What gesture creation in deaf children can tell us about how all children learn language.* New York: Psychology Press.

Gomez, J. C. 2007. Pointing behaviors in apes and human infants: A balanced perspective. *Child Development, 78,* 729–734.

Gowlett, J., C. Gamble, and R. Dunbar. 2012. Human evolution and the archaeology of the social brain. *Current Anthropology, 53,* 693–722.

Gräfenhain, M., T. Behne, M. Carpenter, and M. Tomasello. 2009. Young children's understanding of joint commitments. *Developmental Psychology, 45,* 1430–1443.

Greenberg, J. R., K. Hamann, F. Warneken, and M. Tomasello. 2010. Chimpanzee helping in collaborative and non-collaborative contexts. *Animal Behaviour, 80,* 873–880.

Greenfield, P. M., and E. S. Savage-Rumbaugh. 1990. Grammatical combination in *Pan paniscus:* Processes of learning and invention in the evolution and development of language. In S. T. Parker and K. R. Gibson, eds., *"Language" and intelligence in monkeys and apes* (pp. 540–578). Cambridge: Cambridge University Press.

———. 1991. Imitation, grammatical development, and the invention of protogrammar by an ape. In N. A. Krasnegor, D. M. Rumbaugh, R. L. Schiefelbusch, and M. Studdert-Kennedy, eds., *Biological and behavioral determinants of language development* (pp. 235–258). Hillsdale, NJ: Lawrence Erlbaum.

Grice, H. P. 1957. Meaning. *Philosophical Review, 66,* 377–388.

———. 1975. Logic and conversation. In P. Cole and J. Morgan, eds., *Syntax and semantics,* Vol. 3 (pp. 41–58). New York: Academic Press.

Gundel, J., N. Hedberg, and R. Zacharski. 1993. Cognitive status and the form of referring expressions in discourse. *Language, 69,* 274–307.

Haidt, J. 2012. *The righteous mind.* New York: Pantheon.

Hamann, K., F. Warneken, J. Greenberg, and M. Tomasello. 2011. Collaboration encourages equal sharing in children but not chimpanzees. *Nature, 476,* 328–331.

Hamann, K., F. Warneken, and M. Tomasello. 2012. Children's developing commitments to joint goals. *Child Development, 83*(1), 137–145.

Hampton, R. R. 2001. Rhesus monkeys know when they remember. *Proceedings of the National Academy of Sciences of the United States of America, 98*(9), 5359–5362.

Hare, B. 2001. Can competitive paradigms increase the validity of experiments on primate social cognition. *Animal Cognition, 4,* 269–280.

Hare, B., and M. Tomasello. 2004. Chimpanzees are more skillful in competitive than in cooperative cognitive tasks. *Animal Behaviour, 68,* 571–581.

Hare, B., J. Call, B. Agnetta, and M. Tomasello. 2000. Chimpanzees know what conspecifics do and do not see. *Animal Behaviour, 59,* 771–785.

Hare, B., J. Call, and M. Tomasello. 2001. Do chimpanzees know what conspecifics know? *Animal Behaviour, 61*(1), 139–151.

———. 2006. Chimpanzees deceive a human by hiding. *Cognition, 101,* 495–514.

Harris, P. 1991. The work of the imagination. In A. Whiten, ed., *Natural theories of mind* (pp. 283–304). Oxford: Blackwell.

Haun, D. B. M., and J. Call. 2008. Imitation recognition in great apes. *Current Biology, 18*(7), 288–290.

Haun, D. B. M., and M. Tomasello. 2011. Conformity to peer pressure in preschool children. *Child Development, 82,* 1759–1767.

Hawkes, K. 2003. Grandmothers and the evolution of human longevity. *American Journal of Human Biology, 15,* 380–400.

Hegel, G. W. F. 1807. *Phänomenologie des Geistes.* Bamberg: J. A. Goebhardt.

Herrmann, E., and M. Tomasello. 2012. Human cultural cognition. In J. Mitani, ed., *The evolution of primate societies.* (pp. 701–14). Chicago: University Chicago Press.

Herrmann, E., A. Melis, and M. Tomasello. 2006. Apes' use of iconic cues in the object choice task. *Animal Cognition, 9,* 118–130.

Herrmann, E., A. Misch, and M. Tomasello. Submitted. Uniquely human self-control begins at school age.

Herrmann, E., J. Call, M. Lloreda, B. Hare, and M. Tomasello. 2007. Humans have evolved specialized skills of social cognition: The cultural intelligence hypothesis. *Science, 317,* 1360–1366.

Herrmann, E., M. V. Hernandez-Lloreda, J. Call, B. Hare, and M. Tomasello. 2010. The structure of individual differences in the cognitive abilities of children and chimpanzees. *Psychological Science, 21,* 102–110.

Herrmann, E., V. Wobber, and J. Call. 2008. Great apes' (*Pan troglodytes, Pan paniscus, Gorilla gorilla, Pongo pygmaeus*) understanding of tool functional properties after limited experience. *Journal of Comparative Psychology, 122,* 220–230.

Heyes, C. M. 2005. Imitation by association. In S. Hurley and N. Chater, eds. *Perspectives on imitation: From mirror neurons to memes.* (pp. 51–76). Cambridge, MA: The MIT Press.

Hill, K. 2002. Altruistic cooperation during foraging by the Ache, and the evolved human predisposition to cooperate. *Human Nature, 13*(1), 105–128.

Hill, K., and A. M. Hurtado. 1996. *Ache life history: The ecology and demography of a foraging people.* New York: Aldine de Gruyter.

Hirata, S. 2007. Competitive and cooperative aspects of social intelligence in chimpanzees. *Japanese Journal of Animal Psychology, 57,* 29–40.

Hobson, P. 2004. *The cradle of thought: Exploring the origins of thinking.* London: Pan Books.

Hrdy, S. 2009. *Mothers and others: The evolutionary origins of mutual understanding.* Cambridge, MA: Harvard University Press.

Johnson, M. 1987. *The body in the mind.* Chicago: University of Chicago Press.

Kahneman, D. 2011. *Thinking, fast and slow.* New York: Farrar, Strauss, and Giroux.

Kaminski, J., J. Call, and M. Tomasello. 2008. Chimpanzees know what others know, but not what they believe. *Cognition, 109,* 224–234.

Karmiloff-Smith, A. 1992. *Beyond modularity: A developmental perspective on cognitive science.* Cambridge, MA: The MIT Press.

Kobayashi, H., and S. Kohshima. 2001. Unique morphology of the human eye and its adaptive meaning: Comparative studies on external morphology of the primate eye. *Journal of Human Evolution, 40,* 419–435.

Korsgaard, C. M. 2009. *Self-constitution: Agency, identity, and integrity.* New York: Oxford University Press.

Krachun, C., M. Carpenter, J. Call, and M. Tomasello. 2009. A competitive nonverbal false belief task for children and apes. *Developmental Science, 12,* 521–535.

———. 2010. A new change-of-contents false belief test: Children and chimpanzees compared. *International Journal of Comparative Psychology, 23,* 145–165.

Kuhlmeier, V. A., S. T. Boysen, and K. L. Mukobi. 1999. Scale model comprehension by chimpanzees (*Pan troglodytes*). *Journal of Comparative Psychology, 113,* 396–402.

Kummer, H. 1972. *Primate societies: Group techniques of ecological adaptation.* Chicago: Aldine-Atherton.

Lakoff, G., and M. Johnson. 1979. *Metaphors we live by.* Chicago: University of Chicago Press.

Langacker, R. 1987. *Foundations of cognitive grammar*, Vol. 1. Stanford, CA: Stanford University Press.

———. 2000. A dynamic usage-based model. In M. Barlow and S. Kemmerer, eds., *Usage-based models of language* (pp. 1–64). Stanford, CA: SLI Publications.

Lefebvre, C. 2006. *Creole genesis and the acquisition of grammar*. Cambridge: Cambridge University Press.

Leslie, A. 1987. Pretense and representation: The origins of "theory of mind." *Psychological Review, 94,* 412–426.

Levinson, S. C. 1995. Interactional biases in human thinking. In E. Goody, ed., *Social intelligence and interaction* (pp. 221–260). Cambridge: Cambridge University Press.

———. 2000. *Presumptive meanings: The theory of generalized conversational implicature.* Cambridge, MA: The MIT Press.

———. 2006. On the human interactional engine. In N. Enfield and S. Levinson, eds., *Roots of human sociality* (pp. 39–69). New York: Berg.

Lewis, C. I., and C. H. Langford. 1932. *Symbolic logic.* London: Century.

Lewis, D. 1969. *Convention.* Cambridge, MA: Harvard University Press.

Liddell, S. 2003. *Grammar, gesture, and meaning in American Sign Language.* Cambridge: Cambridge University Press.

Liebal, K., T. Behne, M. Carpenter, and M. Tomasello. 2009. Infants use shared experience to interpret pointing gestures. *Developmental Science, 12,* 264–271.

Liebal, K., J. Call, and M. Tomasello. 2004. The use of gesture sequences by chimpanzees. *American Journal of Primatology, 64,* 377–396.

Liebal, K., M. Carpenter, and M. Tomasello. 2010. Infants' use of shared experience in declarative pointing. *Infancy, 15*(5), 545–556.

———. 2011. Young children's understanding of markedness in nonverbal communication. *Journal of Child Language, 38,* 888–903.

———. 2013. Young children's understanding of cultural common ground. *British Journal of Developmental Psychology, 31*(1), 88–96.

Liszkowski, U., M. Carpenter, T. Striano, and M. Tomasello. 2006. 12- and 18-month-olds point to provide information for others. *Journal of Cognition and Development, 7,* 173–187.

Liszkowski, U., M. Carpenter, and M. Tomasello. 2008. Twelve-month-olds communicate helpfully and appropriately for knowledgeable and ignorant partners. *Cognition, 108,* 732–739.

Liszkowski, U., M. Schäfer, M. Carpenter, and M. Tomasello. 2009. Prelinguistic infants, but not chimpanzees, communicate about absent entities. *Psychological Science, 20,* 654–660.

MacWhinney, B. 1977. Starting points. *Language, 53,* 152–168.

Mandler, J. M. 2012. On the spatial foundations of the conceptual system and its enrichment. *Cognitive Science, 36,* 421–451.

Marín Manrique, H., A. N. Gross, and J. Call. 2010. Great apes select tools on the basis of their rigidity. *Journal of Experimental Psychology: Animal Behavior Processes, 36*(4), 409–422.

Markman, A., and H. Stillwell. 2001. Role-governed categories. *Journal of Experimental and Theoretical Artificial Intelligence, 13,* 329–358.

Maynard Smith, J., and M. Szathmáry. 1995. *Major transitions in evolution.* Oxford: W. H. Freeman Spektrum.

Mead, G. H. 1934. *Mind, self, and society* (ed. C. W. Morris). Chicago: University of Chicago Press.

Melis, A., J. Call, and M. Tomasello. 2006a. Chimpanzees conceal visual and auditory information from others. *Journal of Comparative Psychology, 120,* 154–162.

Melis, A., B. Hare, and M. Tomasello. 2006b. Chimpanzees recruit the best collaborators. *Science, 31,* 1297–1300.

———. 2009. Chimpanzees coordinate in a negotiation game. *Evolution and Human Behavior, 30,* 381–392.

Mendes, N., H. Rakoczy, and J. Call. 2008. Ape metaphysics: Object individuation without language. *Cognition, 106*(2), 730–749.

Mercier, H., and D. Sperber. 2011. Why do humans reason? Arguments for an argumentative theory. *Behavioural and Brain Sciences, 34*(2), 57–74.

Millikan, R. G. 1987. *Language, thought, and other biological categories. New foundations for realism.* Cambridge, MA: The MIT Press.

Mitani, J., J. Call, P. Kappeler, R. Palombit, and J. Silk, eds. 2012. *The evolution of primate societies.* Chicago: University of Chicago Press.

Mithen, S. 1996. *The prehistory of the mind.* New York: Phoenix Books.

Moll, H., and M. Tomasello 2007. Cooperation and human cognition: The Vygotskian intelligence hypothesis. *Philosophical Transactions of the Royal Society of London, Series B: Biological Sciences, 362,* 639–648.

———. 2012. Three-year-olds understand appearance and reality—just not about the same object at the same time. *Developmental Psychology, 48,* 1124–1132.

———. In press. Social cognition in the second year of life. In A. Leslie and T. German, eds., *Handbook of Theory of Mind.* New York: Taylor and Francis.

Moll, H., C. Koring, M. Carpenter, and M. Tomasello. 2006. Infants determine others' focus of attention by pragmatics and exclusion. *Journal of Cognition and Development, 7,* 411–430.

Moll, H., A. Meltzoff, K. Mersch, and M. Tomasello. 2013. Taking versus confronting visual perspectives in preschool children. *Developmental Psychology, 49*(4), 646–654.

Moore, R. In press. Cognizing communicative intent. *Mind and Language.*

Mulcahy, N. J., and J. Call. 2006. Apes save tools for future use. *Science, 312,* 1038–1040.

Muller, M. N., and J. C. Mitani. 2005. Conflict and cooperation in wild chimpanzees. *Advances in the Study of Behavior, 35,* 275–331.

Nagel, T. 1986. *The view from nowhere.* New York: Oxford University Press.

Okrent, M. 2007. *Rational animals: The teleological roots of intentionality.* Athens: Ohio University Press.

Olson, D. 1994. *The world on paper.* Cambridge: Cambridge University Press.

Onishi, K. H., and R. Baillargeon. 2005. Do 15-month-old infants understand false beliefs? *Science, 308,* 255–258.

Peirce, C. S. 1931–1958. *Collected writings* (ed. C. Hartshorne, P. Weiss, and A. W. Burks). 8 vols. Cambridge, MA: Harvard University Press.

Penn, D. C., K. J. Holyoak, and D. J. Povinelli. 2008. Darwin's mistake: Explaining the discontinuity between human and nonhuman minds. *Behavioral and Brain Sciences, 31,* 109–178.

Perner, J. 1991. *Understanding the representational mind.* Cambridge, MA: The MIT Press.

Piaget, J. 1928. Genetic logic and sociology. Reprinted in J. Piaget, *Sociological studies* (ed. L. Smith). New York: Routledge, 1995.

———. 1952. *The origins of intelligence in children.* New York: W.W. Norton.

———. 1971. *Biology and knowledge.* Chicago: University of Chicago Press.

Povinelli, D. 2000. *Folk physics for apes: The chimpanzee's theory of how the world works.* New York: Oxford University Press.

Povinelli, D. J., and D. O'Neill. 2000. Do chimpanzees use their gestures to instruct each other? In S. Baron-Cohen, H. Tager-Flusberg, and D. Cohen, eds., *Understanding other minds: Perspectives from developmental cognitive neuroscience,* 2nd ed. (pp. 111–33). Oxford: Oxford University Press.

Rakoczy, H., and M. Tomasello. 2007. The ontogeny of social ontology: Steps to shared intentionality and status functions. In S. Tsohatzidis, ed., *Intentional acts and institutional facts* (pp. 113–137). Dordrecht: Springer.

Rakoczy, H., F. Warneken, and M. Tomasello. 2008. The sources of normativity: Young children's awareness of the normative structure of games. *Developmental Psychology, 44,* 875–881.

Rekers, Y., D. Haun, and M. Tomasello. 2011. Children, but not chimpanzees, prefer to forage collaboratively. *Current Biology, 21,* 1756–1758.

Richerson, P., and R. Boyd. 2006. *Not by genes alone: How culture transformed human evolution.* Chicago: University of Chicago Press.

Riedl, K., K. Jensen, J. Call, and M. Tomasello. 2012. No third-party punishment in chimpanzees. *Proceedings of the National Academy of Sciences of the United States of America, 109,* 14824–14829.

Rivas, E. 2005. Recent use of signs by chimpanzees (*Pan troglodytes*) in interactions with humans. *Journal of Comparative Psychology, 119*(4), 404–417.

Sandler, W., I. Meir, C. Padden, and M. Aronoff. 2005. The emergence of grammar: Systematic structure in a new language. *Proceedings of the National Academy of Sciences of the United States of America, 102*(7), 2661–2665.

Saussure, F. de. 1916. *Cours de linguistique générale* (ed. Charles Bailey and Albert Séchehaye).

Schelling, T. C. 1960. *The strategy of conflict.* Cambridge, MA: Harvard University Press.

Schmelz, M., J. Call, and M. Tomasello. 2011. Chimpanzees know that others make inferences. *Proceedings of the National Academy of Sciences of the United States of America, 108,* 17284–17289.

Schmidt, M., and M. Tomasello 2012. Young children enforce social norms. *Current Directions in Psychological Science, 21,* 232–236.

Schmidt, M., H. Rakoczy, and M. Tomasello. 2012. Young children enforce social norms selectively depending on the violator's group affiliation. *Cognition, 124,* 325–333.

Schmitt, V., B. Pankau, and J. Fischer. 2012. Old World monkeys compare to apes in the Primate Cognition Test Battery. *PLoS One, 7*(4), e32024.

Searle, J. 1995. *The construction of social reality.* New York: Free Press.

———. 2001. *Rationality in action.* Cambridge, MA: The MIT Press.

Sellars, W. 1963. *Empiricism and the philosophy of mind.* London: Routledge.

Senghas, A., S. Kita, and A. Özyürek. 2004. Children creating core properties of language: Evidence from an emerging sign language in Nicaragua. *Science, 305,* 1779–1782.

Shore, B. 1995. *Culture in mind: cognition, culture, and the problem of meaning.* New York: Oxford University Press.

Skyrms, B. 2004. *The stag hunt and the evolution of sociality.* Cambridge: Cambridge University Press.

Slobin, D. 1985. Crosslinguistic evidence for the language-making capacity. In D. I. Slobin, ed., *The crosslinguistic study of language acquisition,* Vol. 2: *Theoretical issues* (pp. 1157–1260). Hillsdale, NJ: Lawrence Erlbaum.

Smith, J. M., and Eörs Szathmáry (1995). *The Major Transitions in Evolution.* Oxford, England: Oxford University Press.

Southgate, V., C. van Maanen, and G. Csibra. 2007. Infant pointing: Communication to cooperate or communication to learn? *Child Development, 78*(3), 735–774.

Sperber, D. 1994. The modularity of thought and the epidemiology of representations. In L. A. Hirschfeld and S. A. Gelman, eds., *Mapping the mind* (pp. 39–67). Cambridge: Cambridge University Press.

———. 1996, *Explaining culture: A naturalistic approach.* Oxford: Blackwell.

———. 2000. Metarepresentations in an evolutionary perspective. In Dan Sperber, ed., *Metarepresentations: A multidisciplinary perspective.* (pp. 219–34). Oxford: Oxford University Press.

Sperber, D., and D. Wilson. 1996. *Relevance: Communication and cognition.* 2nd ed. Oxford: Basil Blackwell.

Sperber, D., F. Clément, C. Heintz, O. Mascaro, H. Mercier, G. Origgi, and D. Wilson. 2010. Epistemic vigilance. *Mind and Language, 25*(4), 359–393.

Sterelny, K. 2003. *Thought in a hostile world: The evolution of human cognition.* London: Blackwell.

———. 2012. *The evolved apprentice.* Cambridge, MA: The MIT Press.

Stiner, M. C., R. Barkai, and A. Gopher. 2009. Cooperative hunting and meat sharing 400–200 kya at Qesem Cave, Israel. *Proceedings of the National Academy of Sciences of the United States of America, 106*(32), 13207–13212.

Talmy, L. 2003. The representation of spatial structure in spoken and signed language. In K. Emmorey, ed., *Perspectives on classifier constructions in sign language* (pp. 169–196). Mahwah, NJ: Lawrence Erlbaum.

Tanner, J. E., and R. W. Byrne. 1996. Representation of action through iconic gesture in a captive lowland gorilla. *Current Anthropology, 37,* 162–173.

Tennie, C., J. Call, and M. Tomasello. 2009. Ratcheting up the ratchet: On the evolution of cumulative culture. *Philosophical Transactions of the Royal Society of London, Series B: Biological Sciences, 364,* 2405–2415.

Thompson, R. K. R., D. L. Oden, and S. T. Boysen. 1997. Language-naive chimpanzees (*Pan troglodytes*) judge relations between relations in a conceptual matching-to-sample task. *Journal of Experimental Psychology: Animal Behavior Processes, 23,* 31–43.

Tomasello, M. 1992. *First verbs: A case study of early grammatical development.* Cambridge: Cambridge University Press.

———. 1995. Joint attention as social cognition. In C. Moore and P. J. Dunham, eds., *Joint attention: Its origins and role in development.* (pp. 23–47). Hillsdale, NJ: Lawrence Erlbaum.

———. 1998. *The new psychology of language: Cognitive and functional approaches to language structure,* Vol. 1. Mahwah, NJ: Lawrence Erlbaum.

———. 1999. *The cultural origins of human cognition.* Cambridge, MA: Harvard University Press.

———. 2003a. *Constructing a language: A usage-based theory of language acquisition.* Cambridge, MA: Harvard University Press.

———, ed. 2003b. *The new psychology of language: Cognitive and functional approaches to language structure,* Vol. 2. Mahwah, NJ: Lawrence Erlbaum.

———. 2006. Why don't apes point? In N. J. Enfield and S. C. Levinson, eds., *Roots of human sociality* (pp. 506–524). Oxford: Berg.

———. 2008. *Origins of human communication.* Cambridge, MA: The MIT Press.

———. 2009. *Why we cooperate.* Cambridge, MA: The MIT Press.

———. 2011. Human culture in evolutionary perspective. In M. Gelfand, C.-y. Chiu, and Y.-y. Hong, eds., *Advances in culture and psychology,* Vol. 1 (pp. 5–51). New York: Oxford University Press.

Tomasello, M., and J. Call. 1997. *Primate cognition.* Oxford: Oxford University Press.

————. 2004. The role of humans in the cognitive development of apes revisited. *Animal Cognition, 7,* 213–215.

————. 2006. Do chimpanzees know what others see—or only what they are looking at? In S. Hurley and M. Nudds, eds., *Rational animals?* (pp. 371–84). Oxford: Oxford University Press.

Tomasello, M., and M. Carpenter. 2005. *The emergence of social cognition in three young chimpanzees.* Monographs of the Society for Research in Child Development 70(1). Boston: Blackwell.

Tomasello, M. and K. Haberl. 2003. Understanding attention: 12- and 18-month-olds know what's new for other persons. *Developmental Psychology, 39,* 906–912.

Tomasello, M., and K. Hamann. 2012. Collaboration in young children. *Quarterly Journal of Experimental Psychology, 65,* 1–12.

Tomasello, M., and H. Moll. 2013. why don't apes understand false beliefs? In M. Banaji and S. Gelman, eds., *The development of social cognition.* New York: Oxford University Press.

Tomasello, M., S. Savage-Rumbaug, and A. Kruger. 1993. Imitative learning of actions on objects by children, chimpanzees and enculturated chimpanzees. *Child Development, 64,* 1688–1705.

Tomasello, M., J. Call, and A. Gluckman. 1997. The comprehension of novel communicative signs by apes and human children. *Child Development, 68,* 1067–1081.

Tomasello, M., M. Carpenter, J. Call, T. Behne, and H. Moll. 2005. Understanding and sharing intentions: The origins of cultural cognition. *Behavioral and Brain Sciences, 28,* 675–691.

Tomasello, M., M. Carpenter, and U. Lizskowski. 2007a. A new look at infant pointing. *Child Development, 78,* 705–722.

Tomasello, M., B. Hare, H. Lehmann, and J. Call. 2007b. Reliance on head versus eyes in the gaze following of great apes and human infants: The cooperative eye hypothesis. *Journal of Human Evolution, 52,* 314–320.

Tomasello, M., A. Melis, C. Tennie, and E. Herrmann. 2012. Two key steps in the evolution of human cooperation: The interdependence hypothesis. *Current Anthropology, 56,* 1–20.

Tooby, J., and L. Cosmides. 1989. Evolutionary psychology and the generation of culture, part I. *Ethology and Sociobiology, 10,* 29–49.

————. 2013. Evolutionary psychology. *Annual Review of Psychology, 64,* 201–229.

Tuomela, R. 2007. *The philosophy of sociality: The shared point of view.* Oxford: Oxford University Press.

van Schaik, C. P., M. Ancrenaz, G. Borgen, B. Galdikas, C. D. Knott, I. Singleton, A. Suzuki, S. S. Utami, and M. Merrill. 2003. Orangutan cultures and the evolution of material culture. *Science, 299,* 102–105.

Von Uexküll, J. 1921. *Umwelt und innenwelt der tiere.* Berlin: Springer.

Vygotsky, L. 1978. *Mind in society: The development of higher psychological processes* (ed. M. Cole). Cambridge, MA: Harvard University Press.

Warneken, F., and M. Tomasello. 2009. Varieties of altruism in children and chimpanzees. *Trends in Cognitive Science, 13,* 397–402.

Warneken, F., F. Chen, and M. Tomasello. 2006. Cooperative activities in young children and chimpanzees. *Child Development, 77,* 640–663.

Warneken, F., B. Hare, A. Melis, D. Hanus, and M. Tomasello. 2007. Spontaneous altruism by chimpanzees and young children. *PLoS Biology, 5*(7), 414–420.

Warneken, F., M. Gräfenhain, and M. Tomasello. 2012. Collaborative partner or social tool? New evidence for young children's understanding of shared intentions in collaborative activities. *Developmental Science, 15*(1), 54–61.

Watts, D., and J. C. Mitani. 2002. Hunting behavior of chimpanzees at Ngogo, Kibale National Park, Uganda. *International Journal of Primatology, 23,* 1–28.

Whiten, A. 2010. A coming of age for cultural panthropology. In E. Lonsdorf, S. Ross, and T. Matsuzawa, eds., *The mind of the chimpanzee* (pp. 87–100). Chicago: Chicago University Press.

Whiten, A., and R. W. Byrne. 1988. *Machiavellian intelligence: Social expertise and the evolution of intellect in monkeys, apes and humans.* New York: Oxford University Press.

Whiten, A., J. Goodall, W. C. McGrew, T. Nishida, V. Reynolds, Y. Sugiyama, C. E. G. Tutin, R. Wrangham, and C. Boesch. 1999. Cultures in chimpanzees. *Nature, 399,* 682–685.

Wilson, E. O. 2012. *The social conquest of earth.* New York: Liveright.

Wittgenstein, L. 1955. *Philosophical investigations.* Oxford: Basil Blackwell.

Wobber, V., B. Hare, E. Herrmann, R. Wrangham, and M. Tomasello. In press. The evolution of cognitive development in *Homo* and *Pan. Developmental Psychobiology.*

Wyman, E., H. Rakoczy, and M. Tomasello. 2009. Normativity and context in young children's pretend play. *Cognitive Development, 24*(2), 146–155.

# Index